INDUSTRIAL COMPOSTING

Environmental Engineering and Facilities Management

INDUSTRIAL COMPOSTING

Environmental Engineering and Facilities Management

Eliot Epstein

CRC Press
Taylor & Francis Group
Boca Raton London New York

CRC Press is an imprint of the
Taylor & Francis Group, an **informa** business

Cover photo by Jeff Ziegenbein, Deputy Department Manager of Operations and Organics, Inland Empire Utilities Agency, Chino, CA.

CRC Press
Taylor & Francis Group
6000 Broken Sound Parkway NW, Suite 300
Boca Raton, FL 33487-2742

Printed in the United States of America on acid-free paper
10 9 8 7 6 5 4 3 2 1

International Standard Book Number: 978-1-4398-4531-8 (Hardback)

Library of Congress Cataloging-in-Publication Data

Epstein, Eliot, 1929-
Industrial composting : environmental engineering and facilities management / Eliot Epstein.
p. cm.
Includes bibliographical references and index.
ISBN 978-1-4398-4531-8 (alk. paper)
1. Refuse and refuse disposal. 2. Environmental engineering. 3. Facility management. 4. Compost industry. 5. Refuse disposal facilities. I. Title.

TD791.E68 2011
628.4'4--dc22 2010038459

Visit the Taylor & Francis Web site at
http://www.taylorandfrancis.com

and the CRC Press Web site at
http://www.crcpress.com

To my late brother, Elan Epstein

1932–2007

Contents

Preface

Composting of wastes is a very valuable technology in waste management and enhancing our environment. In the past fifteen years, it has evolved into a more sophisticated technology with greater emphasis on environmental and public health aspects. Particular emphasis has been on odor management, volatile organic compounds (VOCs) reduction, and bioaerosols management.

In the early 1970s and 1980s, research in the United States was concentrated on technology development with a special emphasis on pathogens. This resulted in the development of regulations designed to protect public health. In the 1990s considerable research was published on the utilization of compost and its importance in horticulture, erosion control, plant pathology management, and other uses. Today's research is primarily concentrated on emissions and their control.

Because of these successful efforts, composting increased, and more companies entered into the field with new and improved technologies.

In the United States, it is estimated that there are over four thousand composting operations. A majority of these are open facilities composting green wastes. Over 250 facilities are composting biosolids, and, predominantly, these are enclosed or partially enclosed sites. Interest is growing in composting of food wastes. Specialty composting, primarily composting of animal mortalities, has also increased. There is an excellent opportunity for composting of numerous industrial wastes, particularly from pharmaceutical, food processing, paper processing, and other organic waste producers.

From 1955 to 1972, I was at the University of Maine conducting research on runoff and soil erosion, soil–water relations with respect to plants, and management of animal wastes. In addition, I taught soil physics at the University of Maine and guided graduate students. In 1972 I joined a team of excellent scientists at the U.S. Department of Agriculture (USDA), Agricultural Research Station in Beltsville, Maryland.

At the largest USDA agricultural research station in the world, we began studies on land application of biosolids and composting. The principal scientists working on composting and land application of biosolids were Patricia Millner, Rufus Chaney, John Walker, Wiley Burge, George Willson, and Jim Parr. They were the best in their field. Patricia Millner and Rufus Chaney are still at Beltsville and are conducting excellent research.

We began doing research on biosolids composting in 1973. In the period 1973 to 1976, with the assistance of the above-mentioned scientists, I developed the aerated static pile (ASP) method and published the first paper on this subject (Epstein et al., 1976). At that time, we referred to it as the Beltsville method. John Walker began the research on biosolids composting using the windrow method. Two composting machines were available and tested. One was the Cobey Compost turner and the other manufactured by General Motors using the Terex front-end loader. We initially started composting biosolids (treated sewage sludge) from the Blue Plains Wastewater Treatment plant in Washington, D.C. Odor production was not a significant problem,

FIGURE 1 Windrow and ASP systems at Beltsville, Maryland. General Motors Terex composter is shown turning a windrow.

as residents were approximately one mile away, with a very dense, wooded area separating the research facility. At that time, we did not have too much material on our composting pad. Subsequently, undigested (raw sludge) was delivered to the site, and enormous odors were produced. It was then that we developed the ASP system. Figure 1 shows the Beltsville site. Both windrows and the ASP system are shown.

We were very fortunate to have as a neighbor Congresswoman Spellman, who lived in a community north of Beltsville and was concerned about the odors. She was an environmentalist and probably one of the first green congresspersons. With her support, we were able to galvanize public opinion and overcome the odor issue.

Our efforts at Beltsville provided sound scientific data for the U.S. Environmental Protection Agency (USEPA) regulations published as 40CFR503.

In 1980, Joel Alpert and I started E&A Environmental Consultants, Inc. Here, again, in our employment we had excellent engineers and scientists. Individuals such as Todd Williams, Charles Alix, Mark Gould, Larry Sasser, Chris Peiot, Ron Alexander, and Kathy Feldman were leaders in their field and superb scientists and engineers. They were also very decent, ethical individuals. E&A Environmental Consultants, Inc. was an excellent firm dedicated to its clients and the performance of the highest standards in the industry.

I thank Todd Williams, Charles Alix, and John Bouey for their technical advice and help on the three chapters dealing with odors. They provided excellent guidance and suggestions. I also thank Dr. Jonathan Wong, professor at the Baptist University in Hong Kong. He had me involved in two exciting composting projects and arranged for me to teach at the Baptist University of Hong Kong for one month.

Finally, I thank the U.S. Composting Council, its board of directors, members, and especially Dr. Stuart Buckner, for support.

REFERENCES

Epstein, E., Willson, G. B., Burge, W. D., Mullen, D. C., and Enkiri, N. K. 1976. A forced aeration system for composting wastewater sludge. *J. Water Pollut. Control Fed.* 48:688–94.

Acknowledgments

I am most grateful to my wife, Esther, who supported me throughout the year with the preparation of the book and did the editing. She was very patient and provided very valuable suggestions. My children, Beth, Jonathan, and Lisa, strongly encouraged me to proceed with this endeavor.

The Author

Eliot Epstein, Ph.D., is an internationally known expert in composting of waste material. He specializes in composting and biosolids management, solid waste, land application, and waste disposal. He is a soil physicist–agronomist by training. He spent 16 years as Research and Station Leader for the U.S. Department of Agriculture (USDA) Research Service at the University of Maine in Orono, Maine. In 1972, Dr. Epstein transferred to the USDA Research Center at Beltsville, Maryland, where he was the principal scientist on the project that resulted in the development of the "Beltsville" Forced Aeration (Aerated Static Pile – ASP) Composting System for waste.

Dr. Epstein has directed a variety of waste management projects for both government and industry, with tasks ranging from research and analysis to facilities design and operation. He has participated in many major composting programs performed in North America, Europe, and Asia, has authored over 150 technical papers on composting and biosolids-related issues, and has conducted technology transfer programs in sludge management and composting of various wastes for the U.S. Environmental Protection Agency (U.S. EPA) and the U.S. Department of Energy (U.S. DOE).

In 1997, Dr. Epstein authored the book *The Science of Composting* (CRC Press), and in 2002 published a book titled *Land Application of Sewage Sludge and Biosolids* (Lewis Publishers). Dr. Epstein was active in the Water Environment Federation, serving on their Residuals Committee. In 2007, he received the Pioneer Award in Disinfection from the Water Environment Federation. He was project leader on the U.S. EPA project to revise the White House document, 40 CFR 503, on Pathogen and Vector Attraction Control. Dr. Epstein has served as a consultant to the World Bank, the U.S. EPA, and the United National Development Programs.

1 A Prospective

INTRODUCTION

In 1997, I authored a book titled *The Science of Composting*. I wanted the reader to learn about the principal aspects of composting. The knowledge of the basics is extremely important in the management of a composting facility. This knowledge allows the operator to evaluate conditions and to initiate improvements. It also provides the reader with an understanding of aspects related to air quality, public health, regulations, and product use. Some of the material, especially the history of composting in this chapter, I took from *The Science of Composting*.

In this book, *Industrial Composting*, I attempt to provide the reader with the principles related to the design of facilities; considerations for siting; planning; elements of economics and cost; major management issues such as odors, pathogens, and bioaerosols; public relations; and product utilization and marketing.

Composting is the biological decomposition of organic matter under aerobic conditions. This is in contrast to fermentation or anaerobic decomposition that takes place under anaerobic conditions. Furthermore, the definition of composting, particularly as it pertains to large-scale (i.e., industrial) conditions, requires that the process be controlled and managed. What does this involve?

Controlling the process requires that the temperature and oxygen levels be maintained for optimum conditions. The temperature regime involves both mesophilic and thermophilic conditions. Under mesophilic temperatures, biological decomposition of the organic matter is more effective and rapid. Under mesophilic conditions, the rate of decomposition is rapid initially, and the easily decomposable materials, carbohydrates, proteins, and fats, are metabolized by myriad organisms (see Epstein, 1997 for details). Mesophilic conditions toward the end of the process result in a stable and mature product. Under thermophilic conditions, pathogens and weed seeds are destroyed. Many feedstocks contain pathogens, and therefore thermophilic temperatures are necessary for pathogen destruction or inactivation.

Composting is being done for a large number of feedstocks: sewage sludge, biosolids, septage or night soil, manure, animal mortalities, food waste, yard waste, industrial wastes, and military wastes.

Composting is an excellent method of disinfection of wastes. It will destroy bacteria, viruses, and parasites. Consequently, it can play an important role in the disinfection of human and animal wastes in developing countries. If good composting is achieved, i.e., reaching thermophilic temperatures for several days, the compost can be used for crop production, including vegetables.

HISTORY

In the broadest sense, undoubtedly composting was practiced in ancient times. The Israelites, Greeks, and Romans used organic waste directly or composted them (Martin and Gershuny, 1992). One of the earliest uses of manure in agriculture was found on a set of clay tablets in the Akkadian Empire, existing in the Mesopotamian era, 2350 to 2150 B.C. (Martin and Gershuny, 1992). The early civilizations of South America, China, Japan, and India practiced intensive agriculture and use of animal and human waste as fertilizers (Howard, 1943). The Talmud, which is a commentary and elucidation of the Jewish biblical oral law (Mishnah), refers to the use of manure to enrich and moisten the soil. The Talmud also cautions against the use of raw manure and advocates allowing it to rot before use (Martin and Gershuny, 1992). In medieval times two English abbeys, St. Albans (1258) and Priory of Newenham (1388), promoted the use of compost (Martin and Gershuny, 1992). Martin and Gershuny (1992) also refer to the use of compost in the Renaissance literature.

In his book *Five Hundred Pointes of Good Husbandrie*, Thomas Tussler (1909) wrote:

> If a garden require it, now trench it ye may, one trench not a yard, from another go lay;
> Which being well filled with muck by and by, to cover with mould, for a season to lie.

Shakespeare has Hamlet telling his mother: "Confess yourself to heaven, Repent what's past, avoid what is to come, And do not spread the compost on the weeds To make them ranker" (Shakespeare, 1997, Act 3, Scene 4).

Sir Francis Bacon and Sir Walter Raleigh also mentioned composting. The use of compost is well documented in the early American period. Our early presidents George Washington, Thomas Jefferson, and James Madison all advocated the use of manure and compost (Martin and Gershuny, 1992). George Washington was the first great proponent of composting in this country. The father of our nation constructed an unusual building to help in the decomposing of organic matter. He also advocated placing manure and other organic materials in pits for "curing." As he converted his major cash crop from tobacco to wheat, his use of organic compost helped increase crop yield. Interestingly, he was not only thinking of himself. In 1788, he wrote: "Every improvement in husbandry should be gratefully received and peculiarly fostered in this Country, not only as promoting the interest and lessening the labor of the farmer, but as advancing our respectability in a national point of view; for, in the present state of America, our welfare and prosperity depend upon the cultivation of our lands and turning the produce of them to the best advantage" (Washington, 2005). Washington's writings include numerous references to the creation of compost, as well as directions to his foreman on how to make the best compost.

Research on composting in the United States appears to have begun in the 1880s. One of the earliest publications on composting in the United States was Bulletin 61 by the North Carolina Agricultural Experiment Station (1888; Figure 1.1). Maynard (1994) cites seventy years of research on waste composting and utilization at the Connecticut Agricultural Experiment Station. Hyatt (1995) reported that during the period 1971 to 1993 the number of citations, both in the United States and

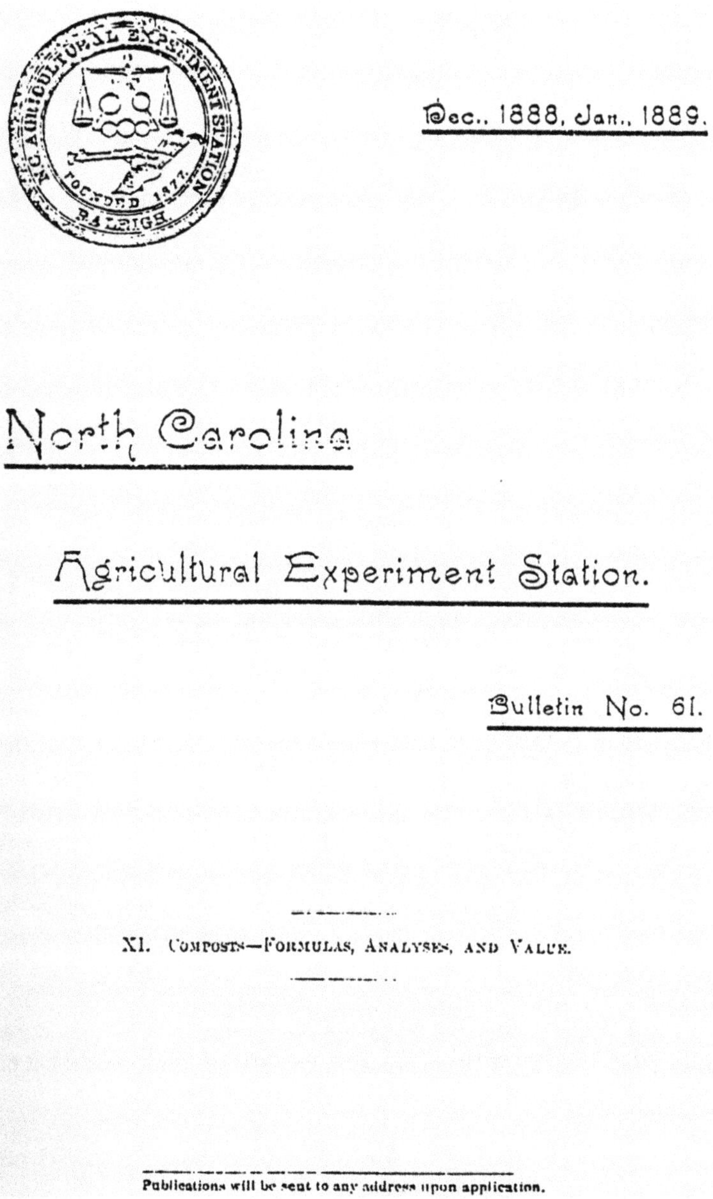

Dec., 1888, Jan., 1889.

North Carolina

Agricultural Experiment Station.

Bulletin No. 61.

XI. Composts—Formulas, Analyses, and Value.

Publications will be sent to any address upon application.

FIGURE 1.1 An early publication by North Carolina Agricultural Experiment Station, Bulletin No. 61, XI. Composts—Formulas, Analyses, and Value. December 1888, January, 1889.

internationally, relating to the subject of compost had grown to 11,353. This indicates strong interest and research effort on the subject.

The concept of large-scale composting in a methodical manner is often attributed to Sir Albert Howard and his Indore process at the Institute of Plant Industry, Indore, Central India, between the years 1924 and 1931 (Howard, 1935). Initially, the process was anaerobic, but later it was modified to an aerobic process and renamed the Bangalore process. The basic concept was to utilize vegetable and animal wastes and night soil (human excrement), mixing them with an alkaline material for neutralizing acidity, and managing the mass through turning for aeration and water addition. The process used either shallow pits or piles that at times contained 909 tonnes (1,000 tons). The dimensions of some compost pits were 9 m (30 feet) by 4 m (14 feet) by 0.9 m (3 feet) with sloping sides. It is interesting to note that Sir Howard (1935) observed "air percolates the fermenting mass to a depth of about 45.7 to 61 cm (18 to 24 inches) only, so for a height of 91 cm (36 inches), extra aeration must be provided." Only in recent years has it been documented that oxygen levels at the bottom of windrows are very limited, and that this zone becomes anaerobic (Epstein, 1997). Van Vuren (1948) published results of composting of urban wastes in 1939 in South Africa based on Sir Howard's principles (Figure 1.2). He viewed the composting of urban wastes as a method of disinfection while producing organic matter, which could restore soil humus. One of the earliest systems was patented by Beccari in 1922 (Beccari, 1922).

The first full-scale refuse composting facility in Europe was established in the Netherlands in 1932. The Van Maanen process, a modification of the Indore

FIGURE 1.2 Composting by the Indore method in South Africa as described by Van Vuren, 1939–1942. (Photograph by Mr. R. Nicholson, Stellenbosch, from Van Vuren, 1949.)

method, was used. The plant was operated by N.V. Vuilafvoer Maatschapii (VAM). Ungrounded refuse was piled in large windrows and turned by overhead cranes (Breidenbach, 1971; Epstein et al., 1976). Following World War II, composting increased in Europe, with practically no composting being conducted in the United States. In the Netherlands two large composting plants were constructed in Mierlo and Wijster using the VAM method. It was estimated that nearly one-third of the Dutch refuse was composted in these two plants. Unit trains from the Hague would bring the waste on to an elevated ramp and dump it, forming large piles. Overhead or mobile cranes would move and turn the waste for several months. Screening would then produce several different grades of compost to be used in agriculture and horticulture. In 1976, the author visited the VAM, Wijster, facility. This method appeared to be well suited for developing countries since it was a relatively low technology utilizing easily available equipment. During a visit to South Yemen on behalf of United Nations Developing Programs, the author recommended that a modification of the VAM system be used in lieu of a high-tech mechanical system originally proposed for that country. This low-technology process was implemented.

One of the foremost advocates and promoter of composting and the use of organic matter in America was J. R. Rodale. He initiated a monthly publication, *Organic Farming and Gardening*, to stimulate the use of compost (Martin and Gershuny, 1992).

Briedenbach (1971) indicated that there were more than thirty composting systems in 1969. Table 1.1 lists several major systems in the world at that time.

The only major research conducted on composting in the 1950s and 1960s was conducted at the Richmond Station of the University of California under the leadership of Drs. Gottas and Goulke.

In the 1960s, the U.S. Public Health Service, forerunner of the U.S. Environmental Protection Agency (USEPA), initiated two major research and demonstration projects on composting municipal solid waste (MSW) with biosolids. One location was at Gainesville, Florida, and the other at Johnson City, Tennessee (Briedenbach, 1971). In addition to process and economic studies, there was considerable research on plant growth and effects on soil by the University of Florida and the Tennessee Valley Authority. The Connecticut Agricultural Experiment Station conducted considerable research on compost utilization in the 1940s, 1950s, and 1970s (Maynard, 1994).

In 1973 the U.S. Department of Agriculture (USDA) at the Beltsville, Maryland, Agricultural Research Center initiated a major research effort on composting of biosolids. The USEPA, Maryland Environmental Service (MES), and Washington, D.C. Council of Governments supported much of the research. During 1975, the USDA research team developed the aerated static pile method (Epstein et al., 1976). The research encompassed process and engineering aspects, pathogen and bioaerosol studies, heavy metal uptake studies, plant growth studies, and microbiological studies. The basis of this research provided the USEPA with the data necessary to formulate the regulations in 40CFR257, which later became the basis for the current regulations under 40CFR503. The University of Maryland's Agronomy and Horticulture departments cooperated with the USDA and conducted independent research on plant growth. This research led to the rapid growth of biosolid composting in the United States.

TABLE 1.1
Municipal Solid Waste Composting Facilities in the United States in 1969

Location	Company	System	Capacity Tons per Day	Waste	Start Date	Status in 1969
Altoona, Pennsylvania	Altoona FAM, Inc.	Fairfield-Hardy	41	MSW, paper	1951	Operating
Boulder, Colorado	Harry Gorby	Windrow	91	MSW	1965	Operating intermittently
Gainesville, Florida	Gainesville Municipal Waste Conversion Authority	Metrowaste conversion	136	MSW, raw biosolids	1968	Operating
Houston, Texas	Metropolitan Waste Conversion Corp.	Metrowaste conversion	327	MSW, raw biosolids	1966	Operating
Houston, Texas	United Compost Services, Inc.	Snell	273	MSW	1966	Closed, 1966
Johnson City, Tennessee	Joint USPHS-TVA	Windrow	47	MSW, digested biosolids	1967	Operating
Largo, Florida	Peninsular Organics, Inc.	Metrowaste conversion	45	MSW, digested biosolids	1963	Closed, 1967
Norman, Oklahoma	International Disposal	Naturizer	32	MSW	1959	Closed, 1967
Mobile, Alabama	City of Mobile	Windrow	273	MSW, digested biosolids	1966	Operating intermittently
New York, New York	Ecology, Inc.	Varro	136	MSW		Under Construction
Phoenix, Arizona	Arizona Biochemical Co.	Dano	273	MSW	1963	Closed, 1965
Sacramento County, California	Dano of America, Inc.	Dano	36	MSW	1956	Closed, 1963
San Fernando, California	International Disposal Corp.	Naturizer	64	MSW	1963	Closed, 1964
San Juan, Puerto Rico	Fairfield Engineering Co.	Fairfield-Hardy	136	MSW	1969	Operating

— continued

TABLE 1.1 (Continued)
Municipal Solid Waste Composting Facilities in the United States in 1969

Location	Company	System	Capacity Tons per Day	Waste	Start Date	Status in 1969
Springfield, Massachusetts	Springfield Organic Fertilizer Co.	Frazer-Eweson	18	MSW	1954 1961	Closed, 1962
St. Petersburg, Florida	Westinghouse Corp.	Naturizer	95	MSW	1966	Operating intermittently
Williamston, Michigan	City of Williamston	Riker	3.6	MSW, raw biosolids corn cobs	1955	Closed, 1962
Wilmington, Ohio	Good Riddance, Inc.	Windrow	18	MSW	1963	Closed, 1965

Source: Breidenbach, A. W., *Composting of Municipal Solid Wastes in the United States*, Rep. Pub. SW-47r, U.S. Environmental Protection Agency, Washington, DC, 1971.

In 1976, Rutgers University in New Jersey initiated studies of biosolids composting at the university and in Camden, New Jersey. Considerable microbiological, process engineering, economics, and utilization research was conducted (Bolan et al., 1980; Kasper and Derr, 1981; Singley et al., 1982). In the late 1970s and early 1980s, when both the USDA and Rutgers began cutting back their research efforts, the University of Ohio began conducting studies on plant growth with an emphasis on composting as related to plant diseases.

A major boost to composting research in the United States occurred following the formation of the Composting Council in 1989 and the generous funding by the Proctor and Gamble Company. Today the United States Composting Council represents the industry.

As a result of the early research emphasizing biosolids composting, major European firms entered the American market in the 1980s, principally as a result of the Clean Water Act of 1972. This federal act provided municipalities with funds to explore and develop innovative technologies. Table 1.2 lists the composting systems in the United States and abroad that became available in the 1970s and 1980s.

In the United States, biosolids and yard waste composting are the two most common feedstocks that are being composted. As a result, there are over 230 biosolids composting facilities and over 3,700 yard waste facilities. MSW composting had a dismal start in the United States. In 1969, there were eighteen facilities, as shown in Table 1.1 (Briedenbach, 1971). From 1969 until the 1980s, there was no growth, and many facilities that operated in 1969 were closed. Today there are fifteen facilities. Although many European systems operated well in Europe, the same systems or others in the United States developed problems.

TABLE 1.2
Composting Systems That Became Available in the United States in the 1970s and 1980s

Aerated Static Pile	Windrow
Dano, United States, Europe	Daneco, United States, Italy
Bedminster, United States	Heidelberg silo, Germany
OTV/OTVD, France, United States	Buhler Miag, United States, Europe
Fairfield Hardy, United States	PLM Selbergs, Sweden
Ebara, Japan	Japan Steel Works
Enadisma, Spain	Recomp, United States
Environment Recovery Systems, United States	BAV, Europe
Ashbook tunnel, United States, Germany	Agripost, United States
Purac, United States	Seerdrum, England, United States
VAM, Holland	Organic Bioconversion, United States
Inge Brikolare, United States	Ag-Renu, United States
Gicom tunnel, United States, Holland	

TABLE 1.3
Amount of MSW Produced and Separately Collected in Europe

Country	Total MSW (metric tonnes)	Organic MSW (metric tonnes)	Separately Collected (metric tonnes)	Separately Collected Percent of Total(%)
Austria	2,800,000	800,000	600,000	75%
Denmark	2,780,000	973,000	652,000	67.01%
Finland	2,510,000	1,004,000	93,000	9.26%
Flanders	3,126,044	1,158,785	723,795	62.46%
France	28,000,000	9,800,000	1,600,000	16.33%
Germany	49,100,000	9,000,000	7,000,000	77.78%
Greece	3,900,000	1,833,000	0	0%
Ireland	2,060,000	556,200	6000	1.08%
Italy	28,400,000	9,542,400	1,500,000	15.72%
Luxemburg	250,000	109,500	34,000	31.05%
Netherlands	8,220,000	3,452,400	1,700,000	50.00%
Portugal	3,800,000	1,406,000	14,000	1.00%
Spain	17,200,000	7,585,200	50,000	0.66%
Sweden	3,810,000	1,500,000	400,000	26.67%
UK	34,000,000	10,880,000	618,517	5.68%

Source: Based on Hogg, D. et al. 2002. *Comparison of compost standards within the EU, North America and Australasia*. The Old Academy, Banbury Oxon, UK.

MSW composting had to compete with landfill tipping fees. Low tipping fees in many areas made it uneconomical to build good composting facilities. As a result, poor facilities were designed, or vendors who had no knowledge of composting saw an opportunity to make money. These vendors often underbid projects, which resulted in low fees and the design of poor facilities or the inability to meet debt payments. Odor problems and poor products resulted and public confidence waned. Multi-million-dollar facilities closed in Portland, Oregon, and in Dade County, Florida, because of poor design and odors. Recently, several major European companies withdrew from the American market. Currently, there is considerable interest and activity in composting of food waste and limited composting of animal wastes.

MSW is by far the largest source of organic waste and potentially could be the largest contributor to composting. Low landfill costs and high costs for producing a clean compost discourage the establishment of MSW composting facilities. In Europe, of MSW or biowaste, separated MSW is more common. Table 1.3 shows data from 2002 on the percentage of separately collected MSW in European countries.

The separation of contaminants from MSW resulting in biowaste would greatly reduce the cost of composting MSW. But until the cost of landfilling is reduced or the public becomes more concerned with greenhouse gases, we will not see a significant increase in MSW composting.

Unfortunately, there is little interest by industry to compost many organic materials such as pharmaceutical, petroleum, large-scale confined animal wastes, food manufacturing wastes, and other organics.

ADVANTAGES AND DISADVANTAGES OF COMPOSTING

Composting has many benefits:

- Many community wastes can be composted. Thus, a single composting facility can handle municipal and industrial organic biosolids, MSW, yard wastes, food wastes, etc.
- A composting facility can be designed and operated to minimize environmental impacts. Odors and bioaerosols can be controlled. With today's technology, facilities and operations should not produce offensive odors.
- Composting can help meet states' landfill reduction and recycling goals.
- Composting can decompose or degrade many organic materials.
- Composting produces a usable product. Not only is the product usable, but as a soil conditioner, it can conserve soil moisture, reduce erosion by improving infiltration, and reducing runoff. These aspects are discussed in detail in the book *The Science of Composting* (Epstein, 1997).
- Compost adds carbon to the soil. It thus sequesters carbon, reducing greenhouse gas.

The major disadvantages to composting are:

- Odor, dust, and bioaerosol emissions can occur during the process. These odors and bioaerosols can be controlled through better facility design and operations management.
- Composting facilities take up more space than some other waste management technologies. Space requirements are often related to storage and market demand.
- A product must be marketed.

THE TECHNOLOGY OF COMPOSTING

This book is designed to provide the reader with the key operational aspects and problems associated with composting. Over my thirty-seven years of involvement in composting, odors have been the single most important operational issue facing composting operations. This issue prevented the establishment of several facilities, closed some facilities, and resulted in legal action against other facilities. Because of odors, citizens were concerned with potential diseases, bioaerosols, or impacts from chemicals. Therefore, I devoted three chapters to odors, as well as a chapter to pathogens and bioaerosols.

It is my hope that owners, operators, managers, workers, and other readers will realize from reading this book that the opportunity, knowledge, and ability exist to operate facilities as good neighbors contributing to ecology. Composting is not only a form of recycling and the reuse of resources, but the highest form of recycling and reuse. We are taking waste that normally would end up in landfills, producing methane, a greenhouse gas, and creating a beneficial product. This product, when used on land, sequesters carbon and reduces greenhouse gases. Composting is an aerobic process and, unlike landfills, does not produce and release methane to the atmosphere.

POTENTIALLY ASSOCIATED RISKS WITH COMPOSTING OR COMPOST USE

The beneficial use of wastes or their products often raises the issue of potential risks. The major risks that have been associated with composting facilities are bioaerosols and odors. Those associated with the product have been heavy metals, pathogens, emerging organic chemicals such as pharmaceuticals and personal care products, and the industrial organics, which are discharged into the wastewater or delivered to the compost site as part of the waste stream. Proper design and operation of composting facilities reduces the dispersion of bioaerosols. Odor control is a function of facility design and management. Composting is extremely effective in eliminating pathogens. There is more recent evidence that many emerging organic chemicals are destroyed by composting. In the past decade, the waste stream that is being composted has become cleaner. Heavy metals in biosolids have been significantly reduced by cleaning the wastewater. U.S. environmental regulations have been a major factor. Although very little municipal solid waste (MSW) composting occurs in the United States, recycling has reduced the contamination of municipal solid waste. Consequently, the compost products produced are much cleaner and safer to use.

Workers are the most exposed individuals. They are exposed more frequently than the public and to higher concentrations of pathogens and bioaerosols. There is no evidence in the United States that workers have become sicker than other municipal workers or the public in general. In Europe, there have been some indications that workers in enclosed MSW composting facilities have become sick from bioaerosols. These aspects are discussed in detail in subsequent chapters of this book.

It is very important to put risks in perspective since we are at all times subject to risks from the food we eat, the water we drink, and the air we breathe. We are also at risk from external sources, such as the automobile we drive, diseases, accidents, and other miscellaneous incidents. In an article written in *Time* magazine on December 4, 2006, Jeffrey Kluger puts the risks we encounter in perspective. Furthermore, he questions "why we worry about the things we shouldn't and more the things we should." In other words, we worry more about probabilities than possibilities. The best example is what happened in South Korea in 2008. Thousands of persons protested when the government decided to permit the resale of beef from the United States. The protests were against mad cow disease. There has never been a single case of mad cow disease in the United States. Yet the Koreans encounter many more possible diseases or risks to their lives. How many of these people eat chicken or eggs contaminated with *Salmonella*? How many do not wash their hands after using toilets? How many ignore good medical practices and die from heart attacks, cancer, and respiratory diseases? To put the dangers and situations we face and encounter into perspective, the following provides information on what occurs in our population annually (CDC, 2004, 2007):

- Total residential deaths: 2,426,264
  - Automobile accidents: 45,316
  - Unintentional falls: 20,823
  - Lightning: 90
  - Unintentional poisoning: 27,531
  - Accident unintentional injuries: 121,599
- Diseases kill 2.43 million people each year
  - Diseases of the heart: 631,636
    - Yet, 20% of all adults still smoke.
    - We consume foods that have trans fats or other cholesterol factors.
    - We eat high-salt foods.
  - Malignant neoplasm: 559,888
  - Cerebrovascular diseases: 137,119
  - Influenza and pneumonia: 56,326
  - Diabetes: 72,449
  - Chronic liver and cirrhosis: 27,555
  - Septicemia: 34,234
  - Anemia: 3,996
  - Malnutrition: 3,153
  - HIV: 12,113
  - *Salmonella*: 34
  - *E. coli* O157:H7: 1

- Bird flu: 0
- Mad cow disease: 0
- Other: 447,805

To put things in perspective:

- Food-related illnesses and death in the United States
  - 76 million illnesses
  - 300,000 hospitalizations
  - 5,000 deaths
  - 2,810 nutritional deficiency deaths
- Number of confined bacterial and parasitic infections
  - *Salmonella*: 6,829
  - *Listeria*: 158
  - *Toxoplasma:* 2,500
  - *Campylobacter*: 5,827
  - *Shigella*: 2,869
  - *Cryptosporidium*: 1,225
  - *E. coli* O157:H7: 547

If you examine these data, you immediately realize that the potential for diseases or illnesses as a result of composting or utilization of the product is nil and insignificant. For example, it was suggested that we should be concerned with mad cow disease since prions may not be biodegraded during the composting project. There has not been a single case of the disease in the United States, and worldwide the cases have been from ingesting beef. Even if prions could exist in the compost, what is the potential for human ingestion and disease? Most people I know do not eat compost. In 2007, there was contamination of spinach by *E. coli* O157:H7 in California resulting from the use of contaminated water and the application of uncomposted manure. Some major producers of vegetables decided not to use compost. Compost properly produced is disinfected and the potential survival of *E. coli* O157:H7 is nil.

Mr. Kluger states: "Our concern regarding risks is that we worry more about probabilities than possibilities. We build barricades against perceived dangers while leaving ourselves exposed to real ones. For example, we dread anything that poses a greater risk for cancer than everyday potential dangers. We feel more comfortable with events, which we believe we can control than with events, which we have little or no control."

However, since a local population will perceive a greater environmental risk if a composting facility is built near them, it is incumbent to design and operate composting facilities with minimal exposures to environmental aspects such as odors, dust, and noise. It is important to communicate and convey to the public the very low danger to health from composting facilities. For example, bioaerosols emissions from composting facilities rarely extend beyond 300 feet (see Chapter 13). Pathogens are destroyed by composting. Odors are a nuisance. They do not represent a health problem (Chapter 9). With our existing technology and knowledge, facilities can be designed and operated with minimal odor and other impacts to the environment.

The use of compost sequesters carbon and reduces carbon dioxide emissions. Composting is the ultimate in recycling. Functionally, the only use of compost is to the benefit of our society. Composting reduces dependence on landfills, incineration, and other nonfriendly environmental technologies. The use of compost improves our soils to reduce runoff and erosion, increases the organic content of the soil for better water utilization by plants, and improves soil structure for better soil aeration for better plant development.

One of the most important aspects of this book is to provide the reader with basic concepts of technology. Another is to provide extensive resources. It is obviously impossible to cover all aspects of composting technology in detail. Therefore, for those readers wanting additional information, numerous references are provided at the end of each chapter.

Relatively few books have been written on composting. The more complete books have been written between ten and twenty years ago. However, there are numerous reports, compendia, proceedings of symposia, and documents on specific subjects. Readers need to discern as to the validity of the written word. In this book, I have attempted to document statements whenever possible through referenced journals. This was not always possible. The Internet is very valuable, but often provides erroneous information. When looking up information on the Internet, the reader should observe who provides the information. If the source is a university, government entity, or medical authority, the information may be relied upon. One of the best resources is the U.S. Composting Council annual meeting, and attending workshops.

Some of the more important resources are:

- *BioCycle* magazine
- *Compost Science & Utilization*
- Haug, R. T., *The Practical Handbook of Compost Engineering*, Lewis Publishers, Boca Raton, FL, 1993
- Epstein, E., *The Science of Composting*, CRC Press LLC, Boca Raton, FL, 1997
- Rynk, R., van de Kamp, M., Willson, G. B., Singley, M. E., Richard, T. L., Kolega, J. L., Gouin, F. R. Laliberty, L., Kay, D., Murphy, D. W., Hoitink, H. A. J., and Briton, W. F., *On-Farm Composting Handbook*, Northeast Regional Agricultural Engineering Service, Ithaca, NY, 1992
- Chiumenti, A., Chiumenti, R., Diaz, L., Savage, G. M., Eggerth, L. L., and Goldstein, N., *Modern Composting Technologies*, BioCycle, Emmaus, PA, 2005

CONCLUSION

Composting is an excellent waste management technology. It can provide excellent disinfection of wastes so that they may be used as a resource rather than disposed of as waste. Implementing composting is much more protective of the environment than landfilling or incineration. Today there is no reason to produce odors or endanger the public or workers. We can design and manage composting facilities in an environmentally accepted manner.

The use of compost in both developing and developed nations can provide humanity with many benefits. The benefit to developing countries is enormous. It can result in destruction of several food-borne diseases, provide nutrients for crop production, conserve moisture in the soil, and improve soil conditions for better crop growth. In developed countries, composting can produce a product that can improve plant growth, reduce runoff and erosion, and minimize landfilling or incineration of waste.

REFERENCES

Beccari, G. 1920. Apparatus for working garbage and refuse of towns. U.S. Patent 1,329,105, January 27, 1920; reissue 15,417, July 25, 1922.

Bolan, M. P., Nieswand, G. H., and Singley, M. E. 1980. *Sludge composting and utilization: Statewide applicability for New Jersey*. New Brunswick, NJ: New Jersey Agricultural Experiment Station, Rutgers University.

Breidenbach, A. W. 1971. *Composting of municipal solid wastes in the United States*. Rep. Pub. SW-47r. Washington, DC: U.S. Environmental Protection Agency.

CDC. 2004. Data and statistics. Centers for Disease Control and Prevention. Atlanta, GA.

CDC. 2007. Data and statistics. Centers for Disease Control and Prevention. Atlanta, GA.

Epstein, E. 1997. *The science of composting*. Boca Raton, FL: CRC Press.

Epstein, E., Willson, G. B., Burge, W. D., Mullen, D. C., and Enkiri, N. K. 1976. A forced aeration system for composting wastewater sludge. *J. Water Pollut. Control Fed.* 48:688–94.

Hogg, D., Barth, J., Favoino, E., Centemero, M., Caimi, V., Amlinger, F., Devliegher, W., Brinton, W., and Antler, S. 2002. *Comparison of compost standards within the EU, North America and Australasia*. The Old Academy, Banbury Oxon, UK

Howard, A. 1935. The manufacture of humus by the Indore process. *J. Royal Soc. Arts* lxxiv:26–60.

Howard, A. 1943. *An agricultural testament*. NY: Oxford University Press.

Hyatt, G. W. 1995. Economic, scientific, and infrastructure basis for using municipal composts in agriculture. In *Agriculture utilization of urban and industrial by-products*. Special Pub. 58. Madison, WS: American Society of Agronomy.

Kasper Jr., V., and Derr, D. D. 1981. *Sludge composting and utilization: An economic analysis of the Camden sludge composting facility*. New Brunswick, NJ: New Jersey Agricultural Experiment Station, Rutgers University.

Kluger, J. 2006. *Time Magazine*. December 4, 2006. Time Inc. New York.

Maynard, A. A. 1994. Seventy years of research on waste composting and utilization at the Connecticut Agricultural Experiment Station. *Compost Sci. Util.* 2(2):13–21.

Martin, D. L., and Gershuny, G. 1992. The *Rodale book of composting*. Emmaus, PA: Rodale Press.

North Carolina Agricultural Experiment Station. 1888. *Composts—Formulas, analyses, and value*. Bulletin 61. North Carolina Agricultural Experiment Station. Raleigh, NC.

Shakespeare, W. 1997. Hamlet. In *The complete works of William Shakespeare*. Wordsworth Royal. Ware, Hertfordshire, U.K.

Singley, M. E., Higgins, A. J., and Frumkin-Rosengaus, M. 1982. *Sludge composting and utilization—Design and operating manual*. New Brunswick, NJ: New Jersey Agricultural Experiment Station, Rutgers University.

Tussler, T. 1909. *Five hundred pointes of good husbandrie*. Dobell Press.

Van Vuren, J. P. J. 1948. *Soil fertility and sewage*. London: Faber and Faber Ltd.

Washington, G. 2005. *Farmer, being an account of his home life and agricultural activities*. Paul Leland Haworth, Rindle. Project Gutenberg. www.gutenberg.org.

2 Basic Concepts of Composting

INTRODUCTION

The effective use of a technology depends on its ability to adhere to the basic concepts of composting. The major factors that affect the rate of decomposition of the organic matter during composting are oxygen and moisture. Many technologies are better at providing and maintaining proper oxygen and moisture. For example, it is more difficult for a windrow system to maintain proper oxygen levels, and it is more difficult for the aerated static pile system to maintain proper moisture levels.

Two other factors affect the composting process: temperature and nutrients, especially carbon and nitrogen.

The rate of decomposition or process limitation is a function of microbial activity. Several conditions can limit or reduce microbial activity, and these in turn influence temperature and the rate of decomposition. These include:

- Low moisture
- Low oxygen content
- Lack of free pore space
- Lack of available carbon or degradable organics

Other conditions that may affect the decomposition process but are much less significant are:

- Available inorganic nutrients
- Presence of toxic substances

My book *The Science of Composting* discusses the basic concepts in much detail. Haug's book *The Practical Handbook of Compost Engineering* (1993) is also a very good source.

TEMPERATURE

Changes in temperature are the result of microbial activity. As shown in Figure 2.1, temperatures rise from ambient to mesophilic and then to thermophilic. During these temperature variations, the microbial population changes. These changes are very profound, as they enable the different microorganisms to metabolize the various components of the feedstocks. Temperature is very important to control pathogens as

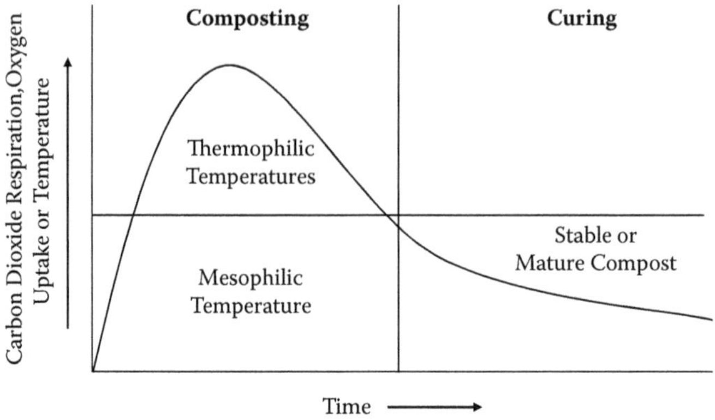

FIGURE 2.1 Changes in temperature and carbon dioxide respiration during composting.

well as to destroy weed seeds. As the process progresses and the available nutrients for the microorganisms are consumed or metabolized, temperatures will drop and at some point return close to ambient. The rate of heat production is proportional to the available organic material for microbial consumption. The decline of temperature to near ambient is an indication that the process is near completion and that the material, probably, is stable and mature.

The type of process and the degree of control have a major effect on temperature. The highest degree of temperature control is usually obtained in static or dynamic enclosed systems. The aerated static pile, for example, provides a higher degree of temperature control than windrows systems. Table 2.1 shows a comparison between an aerated static pile and a windrow for several combinations of yard waste, wood waste, food waste, and mixed waste paper (E&A Environmental Consultants, 1993a). Temperatures in the aerated static pile were consistently higher than in the windrow, and temperatures were lowest with wood wastes since this material has had less available carbon for microbial activity.

There has been some debate regarding the optimum temperature for the decomposition of organic matter. One reason for this controversy is that different feedstocks or materials decompose more rapidly at different temperatures. Most data in the literature indicate that the optimum temperature lies between 50 and 60°C (122 and 140°F). Wiley and Pierce (1955) indicated that the maximum carbon dioxide production occurred at temperatures between 60 and 65°C (140 and 149°F) for mixed garbage and refuse. Schultz (1961) reported that maximum decomposition of municipal solid waste (MSW) occurred at temperatures between 65 and 70°C (149 and 158°F). Other researchers reported that maximum oxygen uptake rates occurred between 45 and 66°C (113 and 150.8°F). Since oxygen uptake is a function of microbial activity, the highest oxygen uptake rate should indicate the most optimum decomposition temperature. However, there are times when one must manipulate the composting process in order to achieve certain goals other than optimum decomposition. This is the case when one needs to maintain temperatures exceeding 55°C (131°F) for several days in order to destroy pathogens. When a feedstock does not contain pathogens or does not need to meet regulatory requirements, lower

TABLE 2.1
The Effect of Composting System and Feedstock on the Number of Consecutive Days at Indicated Temperatures

		Temperature			
Composting System	**Feedstock**	**>50°C (>122°F)**	**>55°C (>122°F)**	**>60°C (>140°F)**	**>70°C (>158°F)**
Aerated	YD/FW	49	30	17	2
static pile	YD/MPW/FW	35	35	32	4
	WW/FW	48	16	7	2
Aerated	YD/FW	34	17	8	5
turned	YD/MPW/FW	26	11	8	4
windrow	WW/FW	13	8	6	4

Source: E&A Environmental Consultants, Inc., *Food Waste Collection and Composting Demonstration Project for City of Seattle Solid Waste Utility*, Final Report, Seattle, 1993a.

Note: YD = yard waste, FW = food waste, MWP = mixed waste paper, WW = wood waste.

temperatures may achieve faster stabilization. When the aerated static pile was first developed for composting of biosolids, disinfection was paramount. Consequently, it was attempted to meet high temperatures exceeding 55°C (131°F) in order to disinfect sludge or biosolids. Once this was achieved, the process was manipulated to reduce temperatures in order to achieve maximum decomposition. When composting feedstocks such as pulp and paper wastes or food processing wastes that do not contain human pathogens, we can achieve faster stabilization at temperatures between 45 and 55°C (118 and 131°F).

Maintaining temperature is often a function of operating conditions. These can include:

- Pile structure
- Pile volume
- Pile insulation
- Pile moisture
- Pile oxygen
- Ambient environmental conditions
- Turning frequency

CARBON AND NITROGEN (C:N)

CARBON

The two most important nutrients for microbial activity and growth that affect the composting process are nitrogen and carbon. A high carbon-to-nitrogen ratio will slow the composting process. A high nitrogen-to-carbon ratio will release ammonia. Although the ideal carbon-to-nitrogen ratio is approximately 27 to 30:1, the composting process is effective within carbon-to-nitrogen ratios of 22 to 40. It can proceed at lower C:N ratios. However, there will be a release of ammonia. At higher C:N ratios, the process slows down.

The carbon provided to the microorganisms on the feedstock is utilized for cellular growth. During microbial metabolism, carbon dioxide is evolved and released to the atmosphere. As the process progresses, the rate of microbial activity decreases and carbon dioxide evolution decreases. It is the relationship between the volatile solids content of a feedstock and carbon dioxide evolution: the higher the volatile solvents content of organic matter, the greater the production of carbon dioxide. Bach et al. (1984) show this in Figure 2.2.

Volatile solids represent the total carbon in a feedstock and not the available carbon. The available carbon is a function of all the chemical constituents of the organic matter. Table 2.2 shows the various components of a feedstock and their susceptibility to mineralization. Starches, sugars, and fats decompose or mineralize much faster than proteins or cellulose, whereas lignin is very resistant to mineralization.

Nitrogen

Microorganisms require nitrogen for cell and protein synthesis. The amount of nitrogen in a waste varies with the type of waste. The nitrogen content and the carbon-to-nitrogen ratio (C:N) are shown in Table 2.3. Microorganisms utilize C and N at a ratio of 30:1. Low carbon-to-nitrogen ratios result in nitrogen volatilization in the form of ammonia and odors. It is important to consider the carbon-to-nitrogen ratio. Many facilities that received large volumes of grass, which has a low C:N ratio,

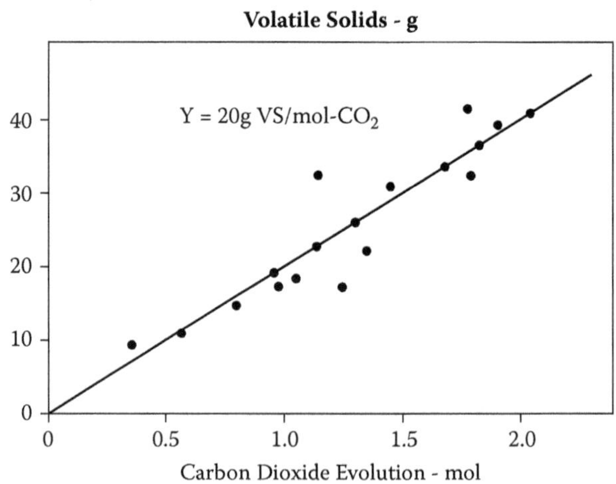

FIGURE 2.2 The relationship between carbon dioxide evolution and volatile solids during the composting of biosolids and rice hulls. (Data from Bach, P. D. et al., *J. Ferment. Technol.*, 62, 285–92, 1984.)

TABLE 2.2
The Susceptibility of the Various Organic Matter Components of the Feedstocks to Decomposition and Mineralization

Organic Matter Components	Susceptibility to Decomposition or Mineralization
Sugars	Very susceptible
Starches, glycogen, pectin	
Fatty acids, glycerol, lipids, fats, phospholipids	
Amino acids	
Nucleic acids	
Protein	Slowly susceptible
Hemicellulose	
Cellulose	
Chitin	
Low molecular weight aromatics and aliphatic compounds	Resistant
Lignocellulose	
Lignin	

encountered odor problems when they did not have a sufficient carbon source in the form of woody waste.

At carbon-to-nitrogen ratios exceeding 50:1, the composting process slows because of rapid cell growth and depletion of available nitrogen. When this occurs, cellular growth is reduced.

MOISTURE

Moisture can be a limiting factor in the composting process. Although we do not have data on the effect of moisture during composting on microbial diversity, we can apply the concepts related to microbial activity in soils. Generally, the rate of microbial activity decreases when the moisture level in compost is below 40%. At 20%, moisture activity is essentially ceased. When moisture content in the compost exceeds 60%, the pore space can be filled with water; then oxygen can become limiting and overall microbial activity decreases. Generally, bacteria are more sensitive to soil moisture than actinomycetes and fungi. Both actinomycetes and fungi tend to predominate in dry soils since they form resisting structures. Since the composting process is a drying process, as water is lost due to the increased temperature, moisture control is essential. The optimal moisture level during the composting process appears to be near 50%. However, the control of moisture is also important from a processing point of view. Many facilities prefer to screen the compost and recover the bulking agent immediately after composting and before curing. Most screens are effective at moisture contents below 45%. Therefore, when possible, maximum composting is carried out at moisture contents between 50 and 55%. Prior to screening, the moisture content

TABLE 2.3
Nitrogen Content and C:N Ratios of Several Feedstocks Used in Composting

Compost Feedstock	Nitrogen Percent Dry Weight	C:N Ratio	Reference
MSW, United States	0.2–3	15	Miscellaneous data
MSW, Japan	1.2–2.7	13–32	Inoko et al., 1979
Digested biosolids	0.5–3.4	15.7	Parker and Sommers, 1983 Poincelot, 1975
Fruit waste	1.52	34.8	Poincelot, 1975
Yard waste	1.95	22.8	Kayhanian and Tchobanoglous, 1992
Paper	0.25	173	Poincelot, 1975
Sawdust	0.11	511	Poincelot, 1975
Grass	2.46–5.0	40–20	Michel et al., 1993 E&A Environmental Consultants, 1994
Leaves	0.93	48	Michel et al., 1993
Produce waste	0.8–2.5	15–25	E&A Environmental Consultants, 1993b
Food waste	3.2	15.6	Kayhanian and Tchobanoglous, 1992 Epstein et al., 2005
Pharmaceutical waste	2.55	19	Poincelot, 1975
Wood (pine)	0.07	723	Poincelot, 1975
Oat straw	1.05	48	Gotaas, 1956
Wheat straw	0.3	128	Gotaas, 1956
Dairy manure	2.9	15	Michel et al., 2004

of the mass is reduced to near 45% for efficient screening. Screening at moisture contents less than 40% can result in dust and release of bioaerosols.

Moisture loss during the composting process is a function of the type of technology, temperature, and ambient conditions. For example, moisture loss is greater with turning systems than with static systems. Turning systems that have aeration such as an agitated bed, will lose more water than nonaerated windrow systems. The higher the temperature in the composting mass, the greater the loss of water.

OXYGEN AND AERATION

Composting is an aerobic process and therefore requires oxygen. Oxygen is provided through aeration. The provision of oxygen depends on the aeration process, which is a function of the system. Windrow composting provides oxygen through the turning process and by convection. Aerated static pile (ASP), agitated bed, and other systems provide oxygen through blowers.

For oxygen to reach the microbial population there needs to be sufficient porosity through the matrix. The porosity is dependent on the feedstock, its moisture content, and particle size. Biosolids and food waste are dense and have high moisture content.

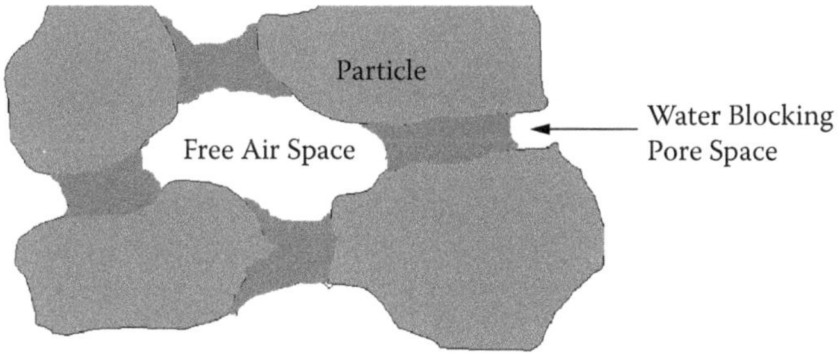

FIGURE 2.3 Free pore space (FPS) in relation to blocked pores with water.

Therefore, bulking agents are usually used. These can either be natural materials, such as wood chips, sawdust, and yard waste, or artificial material, such as shredded rubber tires.

The availability of oxygen is often a function of porosity. Although a source of air (oxygen) can be available, porosity governs its availability to the microorganisms. Total porosity is not an indication of available porosity. Either the free air space (FAS) or pore space and pore size is important (Epstein, 1997) (Figure 2.3).

FAS refers to that portion of the pore space not occupied or blocked by water. The term *FAS* was suggested by Shultz (1961) based on soil aeration concepts of free pore space (Buckingham, 1904). Free air space is that portion of the total pore space that is not occupied with water (Figure 2.3). It is usually calculated by the following equation, which relates the bulk density (BD) to the specific gravity (SG).

$$\text{Free air space (FAS)} = 100\,(1 - \text{BD/SG}) \times \text{dry mass}$$

This pore space allows air to defuse through the media and provide oxygen to the microorganisms. This is illustrated in Figure 2.3, where water blocks some of the pores, reducing the amount of FAS.

Oxygen levels above 10% are usually provided. When the oxygen level is below 5%, it can become limiting to the aerobic microorganisms. At this level anaerobic gases such as methane are generated.

MICROORGANISMS IN COMPOSTING

There are numerous microorganisms involved in the composting process, as shown in Table 2.4. Bacteria are most numerous and are the primary biodegraders. They decompose the more readily available carbonaceous compounds. Details on the microorganisms involved in composting are provided in the book *The Science of Composting* (Epstein, 1997).

STABILITY, MATURITY, AND PHYTOTOXICITY

These terms are often misunderstood and confused. Their understanding is very important, particularly with respect to the quality of compost as related to the process.

TABLE 2.4
Microorganisms Identified in Composting

Bacteria	**Actinomycetes**
Aerobacter (aerogenes)	*Nocardia brasiliensis*
Bacillus megatherium	*Thermomonospora viridis*
B. stearothermophilus	*T. curvata*
B. cereus	*Micromonospora parva*
B. Mycoides	*M. vulgaris*
	Thermoactinomyces vulgaris
Pseudomonad sp.	*Actinoplanes* sp.
(Seven isolates)	*Thermopolyspor polyspora*
Flavobacterium sp.	*Pseudonocardia*
Micrococcus sp.	*Streptomyces violaceoruber*
Sarcina sp.	*S. thermoviolaceus*
Cellumonas folia	*S. rectus*
Chondrococcus exiguus	*S. thermofuscus*
Mycococcus virescens	*S. thermovulgaris*
M. fulvus	*Thermomonospora fusca*
Thibacillus thiooxidans	*T. glaucus*
T. denitrificans	
Proteus sp.	
Fungi	
Rhizopus nigricans	*Absidis orchidis*
Rhizoctonia sp.	*Rhizopus arrhizus*
Geotrichum candidum	*Candida (parapsilosis)*
Mucor pusillus	*Cladosporium herbarum*
Penicillum digitatum	*Rhodotorula rubra*
Mucor racemosus	*Aspergillus tamarii*
Torulopsis sp.	*Zygorhynchus vuilleminii*
Aspergillus flavus	*Trichosporon cutaneum*
Absidia (ramosa)	*Verticillium* sp.
Saccharomyces sp.	*Synecephalastrum* sp.
Pulluloria sp.	*Pichia* sp.
Pythium sp.	*Cylindrocaron* sp.
Hanisenula sp.	*Chaetomium (thermophile)*
Trichoderma koningi	*Lipomyces* sp.
Talaromyces (Penicillium) duponti	*Sporotrichium thermophile*
Stysanus stemonitis	*Fusarium moniliforme*
Glibotrys (alaboviridis)	
Humicola insolens	
Humicola griseus var. *thermoideus*	

TABLE 2.4 (Continued)
Microorganisms Identified in Composting

Protozoans	Algae
Chilomonas (paramecium)	*Hormidium (nitens)*
Cyathomonas (truncata)	*Vaucheria (terrestris)*
Lycogala epidendrum	*Euglena mutabilis*
Cercomonas (crassicanda)	*Protococcus vulgaris*
	Dactylococcus (bicandatus)
	Chlorococcum humicola
	Microcoleus vaginatus
	Porphyridium (cruentum)
	Kentrosphaera sp.
Diatoms (unidentified)	

Stability is a stage in the decomposition of organic matter and is a function of the biological activity. It is a function of the process and is often related to the length of time the process is carried out. The more stable the compost, the lower or slower is the biological activity. In a stable product there is less potential for odors and reheating. An unstable product, when applied to the soil, will rob the soil of the nitrogen that would normally be available to plants.

Maturity is an organic-chemical condition of the compost that indicates the presence or absence of phytotoxic organic acids. This is also related to the process.

Phytotoxicity refers to any substance, organic or inorganic, that is toxic to plants. Phytotoxicity may not be related to the process but could indicate that a chemical is obtained in the feedstock or otherwise included in the process, which can be harmful to plants.

CONCLUSION

Understanding the basic principles of composting can have a significant bearing on the process and product.

Odor production is often the result of a poor understanding of the carbon-to-nitrogen relationship. For example, if a facility is to compost yard waste during the summer months, when grass is deposited in large amounts, it is incumbent on the operator to have an adequate source of carbon. Several facilities have been closed because they did not understand this relationship.

Moisture and oxygen control are also important with respect to process dynamics. They influence odors, process rate, and product quality.

Temperature control and management affects odors, process progress, and product quality.

REFERENCES

Bach, P. D., Shoda, M., and Kubota, H. 1984. Rate of composting of dewatered sewage sludge in continuously mixed isothermal reactor. *J. Ferment. Technol.* 62(3):285–92.

Buckingham, E. 1904. *Contribution to our knowledge of the aeration of soils*. U.S. Department of Agriculture Soils Bulletin 25. Washington, DC: U.S. Department of Agriculture.

E&A Environmental Consultants, Inc. 1993a. *Food waste collection and composting demonstration project for City of Seattle Solid Waste Utility*. Final report. Seattle.

E&A Environmental Consultants, Inc. 1993b. *Composting produce waste and wax coated cardboard using a low technology approach: Pilot study results*. Final report. The Clean Washington Center. Seattle, WA.

E&A Environmental Consultants, Inc. 1994. In-house data.

Epstein, E. 1997. *The science of composting*. Boca Raton, FL: CRC Press.

Epstein, E., Servo, S., Ettlin, L., Cutler, A., and D. Sullivan. 2005. Composting of commercial food waste: Pathogen and related environmental issues. *J. Residuals Sci. Technol.* 2:79–89.

Gotaas, H. B. 1956. *Composting—Sanitary disposal and reclamation of organic wastes*. Geneva, Switzerland: World Health Organization.

Haug, R. T. 1993. *The practical handbook of compost engineering*. Boca Raton, FL: Lewis Publishers.

Inoko, A., Miyamatsu, K., Sugahara, K., and Harda. Y. 1979. On some organic constituents of city refuse composts produced in Japan. *Soil Sci. Plant Nutr.* 25(2):225–34.

Kayhanian, M., and Tchobanoglous, G. 1992. Computation of C/N ratios for various organic fractions. *BioCycle* 33(5):58–60.

Michel Jr., F. C., Adinarayana Reddy, C., and Forney, L. J. 1993. Yard waste composting: Studies using different mixes of leaves and grass in a laboratory scale system. *Compost Sci. Util.* 1(3):85–96.

Michel Jr., F. C., Pecchia, J. A., Rigot, J., and Keener, H. M. 2004. Mass and nutrient losses during composting of dairy manure amended with sawdust or straw. *Compost Sci. Util.* 12(4):323–34.

Parker, C. F., and Sommers, L. E. 1983. Mineralization of nitrogen in sewage sludges. *J. Environ. Qual.* 12:150–56.

Poincelot, R. P. 1975. *The biochemistry and methodology of composting*. Bull. 754. New Haven, CT: Connecticut Agr. Expt. Station.

Schultz, K. L. 1961. *Aerobic decomposition of organic waste materials*. Final report, Project RG-4180 (C5R1). Washington, DC: National Institutes of Health.

Wiley, J. S., and Pierce, G. W. 1955. A preliminary study of high rate composting. *Proc. Am. Soc. Civil Eng.* 81:1–28.

3 The Composting Process

INTRODUCTION

Regardless of the system selected, the elements of the composting process are essentially the same. Configurations vary since each system attempts to optimize the unit processes to achieve the system objectives.

Essentially the composting process consists of three distinct units:

- Preprocessing
- Composting
- Postprocessing

These units will be discussed in detail in subsequent sections of this chapter. However, several other elements need to be considered prior to reviewing the composting process. These are:

- Feedstocks
- Bulking agent or amendments
- Final product distribution and marketing

These elements are very important in the design of a composting facility.

FEEDSTOCKS

The management of feedstocks greatly depends on the type of feedstock and its physical properties. The chemical properties primarily affect the product characteristics. However, some chemical characteristics can influence odor production. Since odor management must consider feedstock handling and storage, the potential impact to feedstock management is explained in Chapter 10.

Numerous feedstocks have been composted. These include:

- Sewage sludge, biosolids, septage, and night soil
- Municipal solid waste, biowaste (source-separated organics)
- Yard waste
- Food waste—grocery, institutional
- Animal waste, fish waste
- Animal mortalities
- Industrial wastes—pharmaceutical, pulp and paper, food processing
- Contaminated soil with hydrocarbons
- Military wastes—HMX, RDX, TNT

The type of feedstock and its physical properties affect the delivery process, storage, and handling of the feedstock prior to the composting process. The primary feedstock characteristics affecting the delivery process are:

- Moisture content or its solids content
- Putrescibility
- Physical properties

The moisture content or solids content determines the type of delivery system employed. For example, biosolids can have a solids content ranging from as low as 16% to over 30%. Biosolids with 16% solids are watery and require a sealed delivery system to prevent spillage on roads and in the facility. Vehicles transporting biosolids need to be covered to contain odors and avoid vectors. These high moisture contents (low solids content) materials would also require different containment storage systems with proper drainage. Yard waste containing very little grass is often transported and delivered in open trucks where approved. Their storage and handling may require chipping prior to incorporation into the composting process.

Biosolids, food waste, and grass are putresable, and their odors must be contained or treated. These materials will also attract vectors such as flies. In addition to being a health issue since vectors can carry and transmit pathogens, they are also a nuisance source. One large biosolids/yard waste facility in California was shut down because of the nuisance source of flies and odors. In Michigan, several facilities handling large volumes of grass were shut down due to odors. In both these cases, proper delivery, storage, and handling as well as an understanding of the basic concepts of composting could have avoided many of the problems.

The physical properties of the feedstock primarily will affect the economics of transportation. Chipped brush or high solids content biosolids are more economical to transport than the same materials as unchipped brush or low solids biosolids. With the high cost of fuel, economic transportation is an important factor in the operation of a facility.

Bulking Agent or Amendments

A bulking agent, as the name applies, is a material generally used to control the moisture content or provide porosity to the feedstock. It can be organic or inorganic to provide structural support to enhance aeration or air movement through the matrix. For example, many of the sewage sludges, biosolids, or manures are in cake form, i.e., mud-like consistency. These materials need to be converted into particle-like material for air to move through the matrix. If the feedstock is very wet, i.e., greater than 65% moisture (35% solids), the addition of sawdust or previously prepared dry compost could be used to reduce the moisture content. The resultant matrix may not provide sufficient porosity. An organic or inorganic bulking agent could then be used to provide the porosity. Wood chips or shredded rubber tire chips would provide porosity. These later bulking agents could be screened out and reused in the process.

Various different bulking agents have been used. The following is a list of the bulking agents used by the author:

- Wood chips
- Wood shavings
- Rice hulls
- Straw
- Bagasse
- Water hyacinths
- Yard waste, tree trimmings
- Shredded rubber chips
- Agricultural wastes
- Pelleted refuse
- Peanut hulls

The use of a bulking agent can be expensive. However, with a little ingenuity, cost can greatly be reduced. Here are some suggestions. Visit the local landfill and see if any material such as chipped brush is being discarded. Often tree companies will deposit material or have some available. Use previously made compost to reduce the moisture content of the feedstock. Try to dry it as much as possible and add it to wet materials such as food waste, manure, biosolids, processing waste, etc. This will also help to control odors. This will reduce the amount of coarse bulking agent needed. Use material that can be screened out and reused, such as hardwood wood chips, shredded rubber tires, and plastics. The bulking agents should be approximately 2.54 to 3.81 cm (1 to 1.5 inches) in size.

An amendment, in contrast, is a material added to either the feedstock or the final product to enhance the process or enhance the final product. If the carbon-to-nitrogen ratio of the feedstock is very high (exceeding 60:1), a nitrogen source may be needed to speed up the process. This could be manure, biosolids, grass, or an inorganic source. An amendment can also be used to enhance the final product in order to increase its value. For example, if the compost has a nitrogen value of 3% or greater, it may be beneficial to add a nitrogen source amendment to increase the product value to over 5%, and therefore qualify as a fertilizer, as well as a soil conditioner.

Bulking agents or amendments are often needed to be able to process the feedstocks in an aerobic and efficient manner. The bulking agents serve several purposes:

- Feedstock moisture content adjustment
- Provide porosity to the matrix
- Adjust the carbon-to-nitrogen ratio
- Product quality
- Stacking (structural stability) and storage

The moisture content and porosity of the matrix are interrelated. Since composting is an aerobic process of decomposition in contrast to anaerobic digestion or decomposition, oxygen through aeration must be provided and available to the aerobic microorganisms.

Final Product Distribution and Marketing

In Chapter 16, I will discuss distribution and marketing of compost. However, it is important to point out how the process affects product quality and hence its marketability. Producing hygienic, stable, and mature compost enhances its marketability and affects the economics of composting. Pathogen destruction and minimizing vector attraction depend on achieving the time-temperature relationships. Regardless of the feedstock, achieving the U.S. Environmental Protection Agency (USEPA) 40CFR503 regulations will result in a hygienic product. A well-cured product will result in a stable and mature product.

THE COMPOSTING PROCESS

The major elements of the composting process are:

- Preprocessing
  1. Feedstock delivery and handling
  2. Feedstock preparation
- Composting
  1. Composting phase or active composting
  2. Curing phase
- Postprocessing
  1. Refining
  2. Product preparation

Preprocessing

Feedstock Delivery and Handling

The delivery and handling of feedstocks depend largely on the type and condition of the feedstock upon arrival at the facility. Putrescent materials need to be handled differently. A great deal depends on location and type of system used. These feedstocks may need to be delivered to an enclosed building with an odor control system to contain odors. This may require a biofilter or other suitable odor management system. For example, food waste, grass, municipal solid waste (MSW), manure, and biosolids delivered to a facility near residences or commercial enterprises may need to be delivered to an enclosed building. This building could be a component of the composting facility. Two illustrations of this type are:

- Davenport, Iowa: Biosolids delivered and deposited into hoppers located within the composting facility, as shown in Figure 3.1. The facility uses the aerated static pile system.
- Edmonton, Canada: MSW is delivered to the composting facility and deposited on a floor in the overall composting facility, as shown in Figure 3.2. The facility uses the Bedminster drum system and an agitated bed system. Curing is outdoors.

FIGURE 3.1 Davenport, Iowa, biosolids and yard waste composting facility. Yard waste to be used as a bulking agent is chipped on site. In the foreground, bags of the finished product are available for sale.

FIGURE 3.2 Edmonton, Canada, municipal solid waste and biosolids composting facility. The rotating drums are in the front of the composting building.

FIGURE 3.3 The LRI indoor windrow composting facility in the state of Washington.

For windrow systems, a separate materials handling building may be appropriate for feedstock delivery. Often, depending on conditions, the feedstock is delivered and deposited directly into the windrow. Rapid mixing will significantly reduce odors. Several facilities have enclosed the feedstock delivery system because of odors to residences nearby. The LRI composting facility in the state of Washington is a totally enclosed windrow system. This is shown in Figure 3.3. The system uses a turning machine with an elevated conveyer and a side chute for building the windrows. Notice the large amount of steam due to the heat. A good ventilation system is needed to remove the high amount of vapor, odors, and dust.

Feedstock Preparation

The objectives of feedstock preparation can be:

- Particle size reduction: The smaller the particle of the material to be composted, generally the faster the decomposition rate. Most of the microbial decomposition occurs on particle surfaces. The smaller the particles, the greater the surface area available for microbial decomposition. It must be kept in mind that particle size is also important from a structural point of view (Haug, 1993). Grinders and shredders (Chiumenti et al., 2005) can achieve this.
- Removal of undesirable material: With heterogeneous feedstocks such as MSW, it is often desirable to remove ferrous and nonferrous metals, plastics, rocks, and other material. Some technologies, such as DANO and Bedminster, prefer to remove material primarily after curing. Most systems conduct contamination removal prior to the composting process, and then further refine in the postcomposting phase. One of the issues

has been whether leaving contaminants such as metals in the system during composting results in a higher level of contamination. As Chiumenti et al. (2005) point out, the separation of contaminants is more difficult when the feedstocks are wet, and therefore it is more desirable to remove contaminants after curing. However, incoming MSW is relatively dry and contaminants can be separated prior to composting. Numerous technologies for separation of various materials are available. A partial list is indicated below:

- Screening: Removal of large plastics, cardboard, large paper, metal fragments.
- Hand sorting: Recyclables, plastics, cardboard, and miscellaneous items.
- Magnetic separation: Ferrous metals.
- Eddy current: Aluminum separated prior to composting.
- Air classification: Paper, plastic light material, glass, metals, miscellaneous materials, both light and heavy.
- Wet separation: Metals, glass, stone, miscellaneous materials, both light and heavy.
- Ballistic separation: Light and heavy items such as plastics, metals, glass, gravel.

Earlier studies reported on the level of heavy metals as a result of various separation strategies (Richard and Woodbury, 1992). However, there have been considerable changes in our waste streams. For example, inks used in printing in the 1990s and earlier contained heavy metals. Cadmium ink gave paper its yellow color. Mercury, zinc, and nickel were important contributors to the U.S. MSW waste stream (Richard and Woodbury, 1992). Heavy metals are not used in inks today, and colors are predominantly from organic dyes. There are changes in the use of mercury, and much less is discarded today.

The heavy metal content in sewage sludge and biosolids has dramatically been reduced because of pretreatment (Chaney et al., 2001). Furthermore, plastic pipes in home and commercial construction have replaced lead pipes as well as copper pipes. This has reduced these elements in the water and in the resulting wastewater and sludge.

Mixing: Mixing is often necessary to obtain a homogeneous matrix for proper composting. This may occur when a bulking agent is added to a feedstock and nonturning technologies are used. Mixing is also used if an amendment is added prior to composting. As Chiumenti et al. (2005) point out, material collected at curbside in bags may need to be mixed with other types of biomass to produce a homogenous mixture. Anytime two or more materials are to be used in the composting process, it is often best to mix these materials prior to composting.

Composting

Composting Phase or Active Composting

The composting phase as indicated here is often referred to as active composting. This is to distinguish this phase from curing. The objective of the composting process (composting and curing phases) is to produce a hygienic, usable product. In this way, the product will have an economic value and be accepted for marketing by the public. To meet this objective, the composting process must:

- Control human and animal pathogens
- Control plant pathogens
- Minimize vector attraction
- Destroy weed seeds
- Prevent the regrowth or reestablishment of pathogens
- Destroy volatile organic compounds (VOCs), which can produce odors

The principal criterion needed to meet these objectives is time-temperature. The other variables, such as moisture, aeration, and carbon-to-nitrogen ratio, impact the process, and this in turn affects the temperature regime. Temperature affects the microbiological process, as well as the rate and extent of decomposition. This time-temperature relationship, as it affects pathogens, vector attraction, and odors, will be discussed in subsequent chapters.

Process configurations vary with the feedstock, its contamination, and the utilization of the final product. The more homogeneous the feedstock, e.g., biosolids or pharmaceutical wastes, the fewer the number of unit processes.

Figure 3.4 illustrates a generic process flow for most feedstocks. Feedstock preparation involves the addition of bulking agent. For example, after delivery of biosolids to an outdoor composting facility by trucks, it can be deposited into a three-sided concrete bin. During feedstock preparation, a front-end loader (FEL) will pick up the biosolids and deposit the contents into a mixer, as illustrated in Figure 3.5. Wood chips or ground yard waste will be added to the mixer in an appropriate ratio. The mix will then be conveyed to an aeration system for composting.

In a windrow operation, the biosolids and bulking agent are placed in the windrow, and the windrow machine does the mixing. This is not a preferred method, as it can result in significant odors. Whenever possible, especially with a putresable feedstock, mixing in a building is preferable.

In the case of biosolids, when a clean, uniform bulking agent, such as wood chips of yard wastes is used, the compost refining step usually consists of screening. There are two options. In the first option, screening is done prior to curing. In this case, less material needs to be cured. In the second option, screening is done after curing. Curing is carried out with the bulking agent in the matrix. However, under this later option, the bulking agent will be more decomposed. If the intention is to recycle as much of the bulking agent as possible, it is best to remove it prior to curing. If screening is done after curing, more space is needed. If the entire process is in a building, the building size to accommodate the curing is much bigger, and the odor control system is bigger.

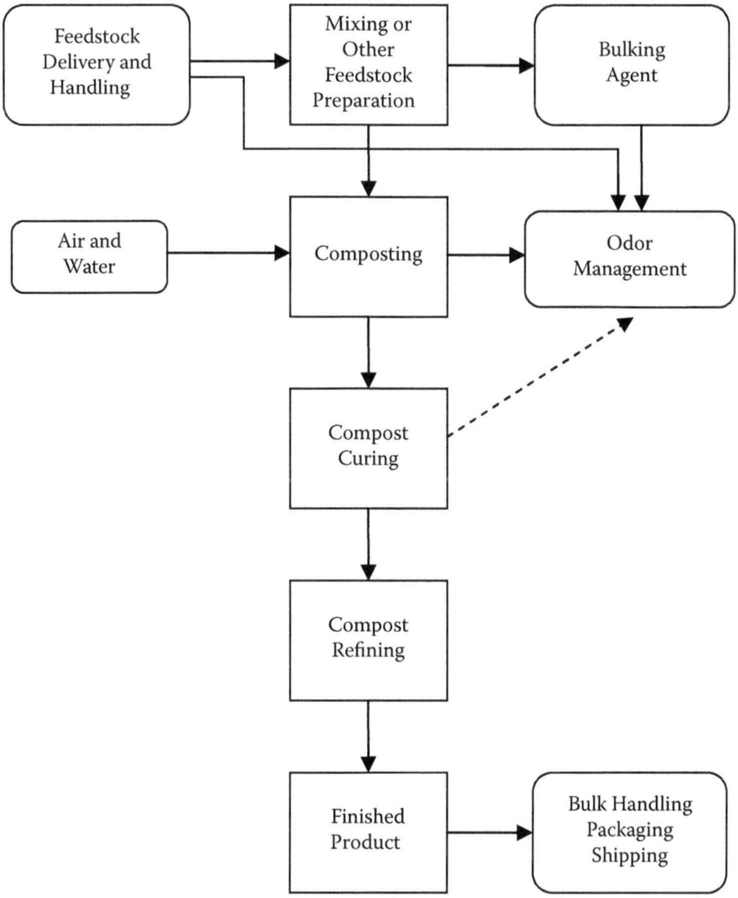

FIGURE 3.4 Basic composting process flow.

Figure 3.6 illustrates a typical process flow for a within-vessel sludge or biosolids system. There could be several options within this process flow:

- The hopper and mixer could be a single unit or separate units.
- Curing could occur before or after screening. This author prefers curing after screening, as it reduces the amount of space needed for curing, and the bulking agent does not deteriorate during curing.
- Composting can be achieved by either static or agitated systems.
- Drying may be a necessity. Therefore, it is indicated as an option.
- Within this system, there are several options for material handling. These include front-end loaders and conveyors.

MSW, because of its heterogeneity and contamination, requires the most configurations. Figure 3.7 illustrates one type of MSW configuration. As indicated earlier, some drum systems do not do any extensive separation in the beginning except

FIGURE 3.5 Stationary mixer depositing biosolids and wood chips into a three-sided concrete bin.

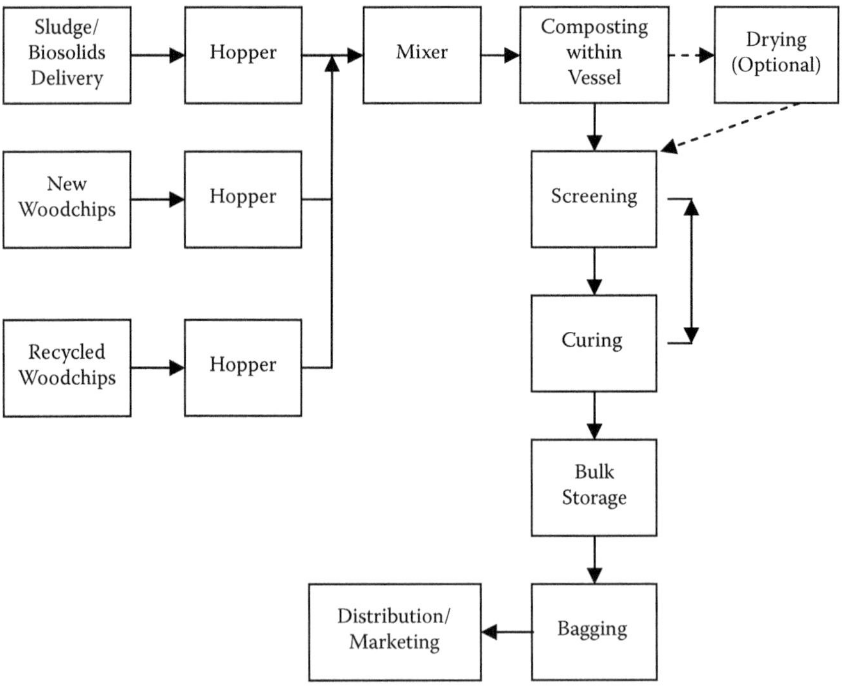

FIGURE 3.6 Process flow illustration for sludge or biosolids composting within vessel.

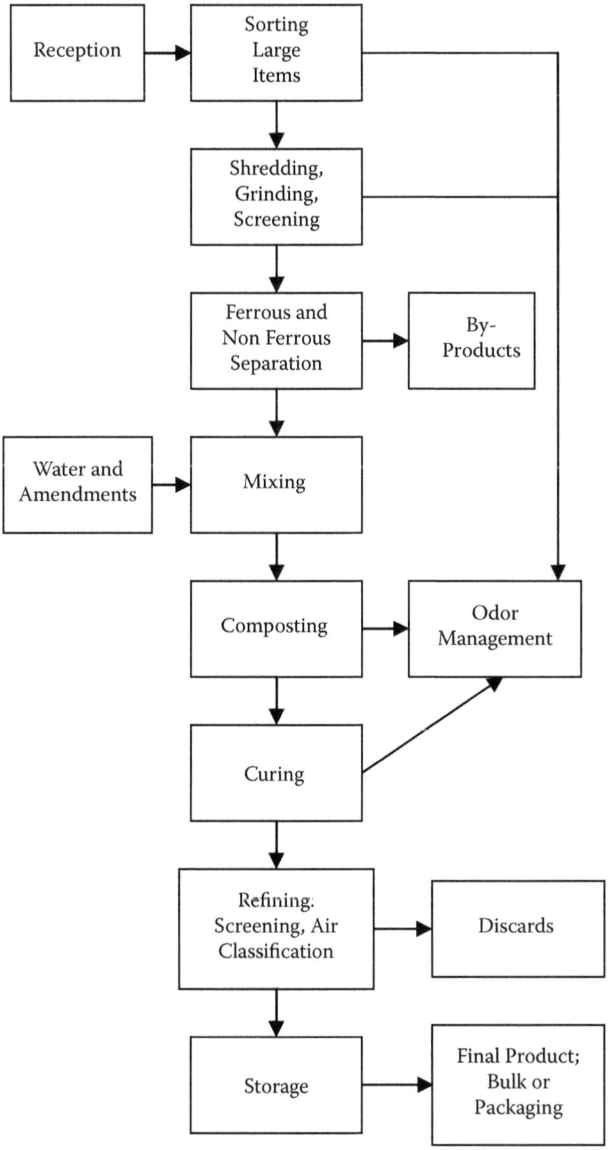

FIGURE 3.7 Process flow for some MSW facilities.

removal of bulky items, which occurs on the tipping floor. Other systems, especially in the past, have resorted to hand separation alongside a conveyor as the MSW flows to the composting system. Today many systems (other than drum) conduct extensive separation prior to composting. This separation primarily attempts to remove ferrous metals, aluminum, and other nonmagnetic material, glass, and plastics. In

Edmonton, Canada, the MSW is processed with biosolids in drums. The refining occurs after the composting process but prior to curing.

Another configuration of a multiple feedstock system is illustrated in Figure 3.8. This system was designed in Hong Kong to provide the agency with the option of shifting to various feedstocks, such as animal waste or food waste. This is somewhat similar to several other MSW composting facilities. In the Ngau Tam Mei horse manure composting plant in Hong Kong, the manure is brought in plastic bags. The manure is removed from the bags inside a receiving building and a bobcat places the manure into a hopper. Using a conveyor system, the manure passes under a magnet to remove ferrous metals. Then it is conveyed through an eddy current system to remove aluminum and other nonmagnetic metals. It is then conveyed onto a disc screen to remove plastics. Each one of the undesirable materials is deposited into containers. Some of these materials can be recycled. The clean manure is then conveyed into a mixer/shredder, as it contains straw. Water and amendments can be added to adjust the moisture content or the carbon-to-nitrogen ratio as needed.

The mixed and shredded material is then conveyed to a drum for initial composting. The retention time in the drums depends on their capacity and the volume to be processed. Since the drums do not complete the composting process, the material from the drums is composted under aeration in covered bins. Similarly, curing occurs in aerated covered bins. When the compost reaches a desired stage in stability and maturity, it is then moved to a storage area where screening is done. Screen size depends on the market requirement. The final product can then be marketed in bags or in bulk.

Curing Phase

A major part of the composting process is curing. The major purpose of curing is product stabilization. Curing can be done before screening or after screening. If curing is done after screening, as shown in Figure 3.6, the bulking agent is removed. This avoids deterioration of the bulking agent and more is recovered. Less space is needed for curing. Under these conditions forced aeration is best.

If the compost is screened after curing, more space is needed for curing. With the bulking agent still in the compost, convective air may be sufficient for oxygen and forced air may be unnecessary. The bulking agent will deteriorate, reducing its recycling potential.

During this stage the compost can achieve stability and maturity. Stability is a function of the process and denotes that sufficient biological decomposition has occurred. When a product is stable, oxygen consumption and carbon dioxide evolution by the microbial community are low. The product will have an earthy odor and not be offensive. A mature product is one where fatty acids are low, and the product would not affect plant growth.

Curing can be done under forced aeration or under natural aeration. Forced aeration at very low airflows has been demonstrated to excel the process. In doing so, less time is required, and therefore less space.

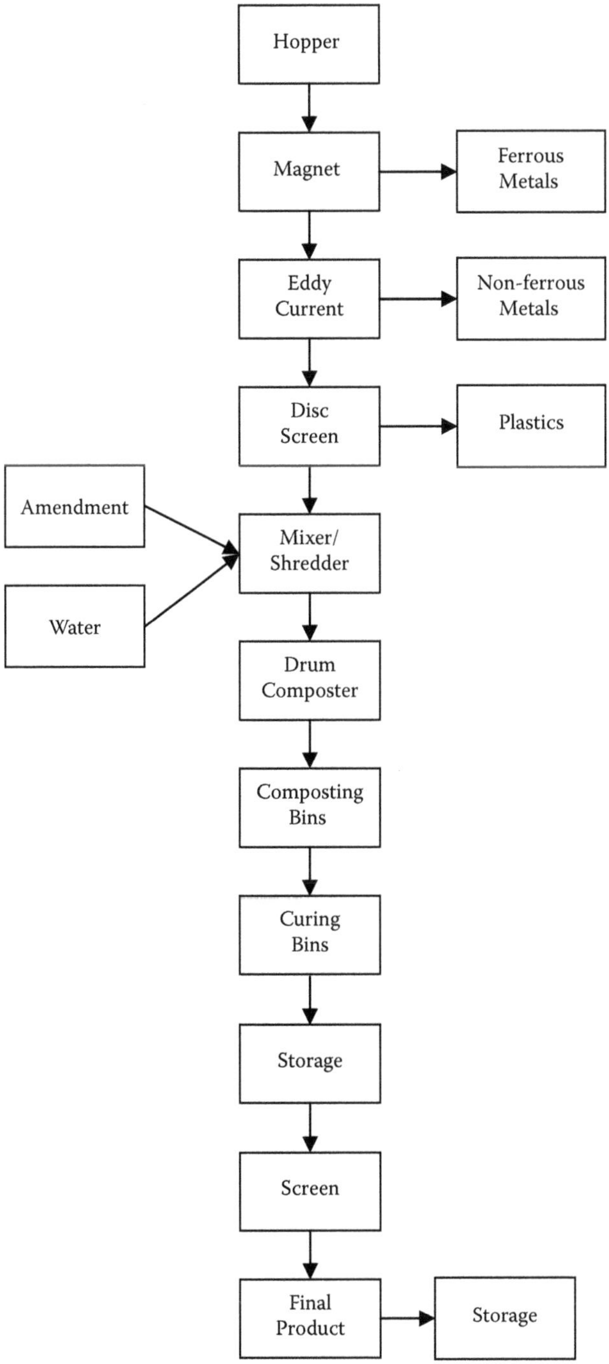

FIGURE 3.8 Multiple-purpose composting system flow process for several different types of feedstocks.

Postprocessing

Refining

Postprocessing is the final phase in the overall composting process. The objective of this phase is to refine the product primarily from a physical perspective. The chemical nature of the product will generally not change from the earlier curing phase. The exception to this statement is if the product is chemically or biologically enhanced to increase its marketability.

The major elements of this phase are usually screening and air classification. During screening, particle size is reduced to from 3 to 9.5 mm (0.125 to 0.375 inch) unless the product is primarily going to be used as mulch. In that case, the particle size is usually larger. The very small particle size, 3 mm (0.125 inch) is often used for golf courses and turf. The 9.5 mm (3/8 inch) particle size is the most widely used. Screening is best achieved when the compost has a moisture content of 40 to 45%. At higher moisture contents, it is difficult to properly screen. At moisture contents below 40%, the material is dusty.

Another aspect of the postprocessing phase is refining. Refining is a density and size separation process. It typically removes glass, metals, wood, film plastic, hard plastic, and other physical contaminants. The term often used is *man-made inerts*. Air classifiers can achieve separation along with magnets, eddy current classifiers, and other equipment. Moisture is again critical. Below 40%, excessive dust occurs. Details of this process are in Chiumenti et al. (2005).

Product Preparation

Screening and refining produce a product for its ultimate use. The better the final product, the greater are its potential uses. Most compost products are the result of screening through a 9.5 mm (3/8 inch) screen. This product has a wide range of uses in horticulture, parks, public works, land reclamation, and other projects. If the product is to be used for turf cultivation, a finer product is needed. For use as mulch, a coarse product is desirable. More details are in Chapter 16.

Typically compost is shipped in bulk. The economical range of transportation is usually considered within 80 km (50 miles). Bagging the product increases the range and usually brings in greater remuneration for the product.

CONCLUSION

The composting process consists of three major elements:

- Preprocessing
- Composting including curing
- Postprocessing

The various process units within each of these elements will vary depending on the feedstock and its nature, the composting system, and the product desired.

Preprocessing primarily involves preparation of the feedstock in order to achieve maximum biological decomposition of the organic matter.

The objective of composting and curing is to obtain a hygienic, stable, and mature product, as well as a product that will enhance the soil physical properties for optimal plant growth.

Postprocessing consists of refining the product. It involves physical size preparation and removal of man-made contaminants. The chemical nature of the product is primarily a function of the chemicals in the feedstock and any additional chemicals added during the process.

REFERENCES

Buckingham, E. 1904. *Contribution to our knowledge of the aeration of soils.* U.S. Department of Agriculture Soils Bulletin 25. Washington, DC: U.S. Department of Agriculture.

Chaney, R. L., Ryan, J. A., Kukier, U., Brown, S. L., Siebielec, G., Malik, M., and Angle, J. S. 2001. Heavy metal aspects of compost use. *In Compost utilization in horticultural cropping systems*, ed. P. J. Stoffella and B. A. Kahn. Boca Raton, FL: Lewis Pub., CRC Press.

Chiumenti, A., Chiumenti, R., Diaz, L., Savage, G. M., Eggerth, L. I., and Goldstein, N. 2005. *Modern composting technologies.* Emmaus, PA: BioCycle.

Epstein, E. 1997. *The science of composting.* Boca Raton, FL: CRC Press LLC.

Haug, R. T. 1993. *The practical handbook of compost engineering.* Boca Raton, FL: Lewis Publishers.

Richard, T. L., and Woodbury, P. B. 1992. The impact of separation on heavy metal contaminants in municipal solid waste composts. *Biomass Bioenergy* 3(3–4):195–211.

Schultz, K. L. 1961. *Aerobic decomposition of organic waste materials.* RG-4180 (C5R1). Washington, DC: National Institutes of Health.

4 Design and Material, Energy, and Water Balances

INTRODUCTION

The objective of this chapter is to provide the reader with basic elements, concepts, and the basis for the design of composting facilities. One of the most useful tools is technical memoranda (TMs), which are related to every phase of the design. The more comprehensive these memoranda are, the better is the understanding of how the facility should be built and operated. The memoranda are usually between the engineers, architects, and others involved in the design and the client. One of the most important aspects of these memoranda is communication. It gives the client the opportunity to question the designers as to the rationale for including specifics in the design. It affords the opportunities to evaluate alternatives.

In this chapter, material balances will be discussed. The information produced by a material balance is very important in design. It provides information that could be used in the sizing, planning of a facility, and product marketing.

Once a site is selected and information is available as to the quantity of waste that would be available so that a material balance can be conducted, one can proceed to a design of the facility. Before conducting the design, the specific system or technology needs to be selected.

Initial TMs and workshops can be used to discuss with the client the potential technologies that may be appropriate. Most often, technical memoranda become important during the design phases. The more complicated or involved the facility is, the more detail and importance the technical memoranda and client interaction have.

Three processes govern the composting process: material, energy, and water balances. Material balance is the most important one. It provides information on the quantity of materials to be used. This intern provides necessary information on site size, size of components such as composting, curing, and storage, the size of structures, and the size and number of equipment. It indicates the quantities of material that need to be handled. This information is useful to identify the equipment size needed. Furthermore, it indicates the quantity of compost to be produced that needs to be marketed.

Energy balances provide information on heat output. From a design point of view, this information can be useful for assessment of equipment sizes and economics. This is also important for pathogen destruction and water removal from the process.

Heat produced will be lost from the piles (the term *pile* here refers to static or turned piles/windrows) to the atmosphere. To some extent the outside temperature, pile insulation, and management of the pile or windrow govern this. Turning results in heat loss and pile cooling. It is best to contain the composting process temperatures to no greater than 65°C (150°F). However, at 55°C (131°F), pathogens are destroyed after several days. Temperatures in the range of 55 to 65°C (131 to 150°F) will indicate that throughout the pile or windrow all of the composting material is subject to temperature sufficient to destroy pathogens.

The water balance indicates the water contained in the feedstocks, the mixture, and the final product. The most important is the mixture water content. This must be less than 60% moisture or 40% solids. Some systems are able to utilize mixtures greater than 60% moisture. At moisture content greater than 60%, there is less free pore space necessary for proper flow of air and provision of oxygen necessary for microbial activity. Water loss from the media is a function of heat and is lost as vapor. The addition of water can be difficult, depending on the process. In a windrow system or agitated bed, water can be added during the turning process. It is more difficult to control in an aerated static pile (ASP) system. Therefore, maintaining the correct moisture depends largely on the mixture moisture content and temperature control. In ASP blowers control moisture and temperature. Often this is done using timers that regulate the on-off cycle of the blower. Moisture control is also needed to screen or refine the final product. Screens and air classifiers operate best at low moisture contents. However, it is not advisable to have the final product too dry, as its handling results in dust, which can be a health hazard to employees or can result in the dispersion of bioaerosols. The ideal moisture content of the final product is between 40 and 45%. The water balance will indicate water needs for proper composting, dust control, and to some extent, odor control. It can provide information on potential for storm water management. This is part of the facility design.

The reader will be provided with information on these three balances with emphasis on material balances. The reader will not be provided with mathematical calculations. These can be found in texts, as indicated below:

Haug, R. T., *The Practical Handbook of Compost Engineering*, Boca Raton, FL, Lewis Publishers, CRC Press, 1993

Rynk, R., van de Kamp, M., Willson, G. B., Singley, M. E., Richard, T. L., Kolega, J. L., Gouin, F. R., Laliberty, L., Kay, D., Murphy, D. W., Hoitink, H. A. J., and Briton, W. F., *On-Farm Composting Handbook*, Ithaca, NY, Northeast Regional Agricultural Engineering Service, 1992

APPROACHES TO DESIGN

Two approaches to design will be discussed: technical memoranda and workshops. Both of these are intended to foster communication between the designer and the client. If the relationship is on a one-to-one basis, then these are often held to a minimum. However, in the case of public entities, often considerable communication must be imparted for final decision making with the involvement of several agencies.

However, even when the decision making is up to an individual or partners, there is often the need for communication with permitting agencies. This is where TMs and workshops may be needed. After the TMs and workshops are completed, a design basis report is usually prepared. This report essentially summarizes the consensus for the design.

Technical Memoranda

TMs may be relatively simple or very detailed, depending on the complexity of the design, system selected, and operations. If the facility needs structures, then the design could be more complex. These TMs could include electrical aspects, aeration, odor and dust control, structural issues, etc. TMs should provide information on equipment and their capabilities. TMs could also be provided on operations indicating responsibilities, contact with regulatory agencies, public information and relations, and other aspects. Details depend on the specific conditions. The following are several examples:

Example 1: Outline for TM on staffing analysis

- I. Staffing requirements
  - A. Factors affecting staffing levels
  - B. Estimated staffing requirements
- II. Staffing responsibilities
  - A. Management
  - B. Operations
  - C. Worker safety and health
  - D. Public relations and regulatory aspects
- III. Compost facility staff functions
  - A. Management and supervision
    1. Office management
    2. Procurement
    3. Accounting and billing
  - B. Direct production labor
    1. Feedstock receiving and handling
    2. Feedstock mixing and preparation including bulking agent
    3. Composting operations
    4. Curing operations
    5. Product preparation including screening, storage, and bagging (if appropriate)
    6. Site maintenance
  - C. Equipment and infrastructure maintenance
  - D. Sampling and analysis
  - E. Support services
  - F. Marketing and sales

Example 2: Outline for TM for odor and dust control

- I. Odor control

A. Significance of problem
    1. Compounds associated with biosolids composting
    2. Odor emissions as related to processing sequence
    3. Odors as nuisance or a health issue
B. Odor sources
    1. Feedstock delivery and storage
    2. Building air
        a. Mixing area
        b. Composting area—surface emissions
        c. Curing area—surface emissions
        d. Screening
    3. Product storage
    4. Negative aeration exhaust
    5. Biofilter
    6. Traffic areas and housekeeping
C. Methods of assessing odors
    1. Baseline odor modeling
    2. Odor collection and analysis
D. Factors affecting odor emissions
    1. Agitation
    2. Temperature
    3. Moisture
    4. Aeration—positive and negative
    5. Materials handling
E. Regulatory requirements and permits
    1. State and county regulations
    2. Occupational regulations (OSHA)
    3. City and local regulations (including fire department)
    4. Permits required
F. Ventilation code analysis
    1. Review of existing standards
    2. Air exchange rate summary
G. Alternative control systems
    1. Biofilters
        a. Compost filters
        b. Sand filters
    2. Scrubbers
    3. Biotrickling filters
    4. Comparative matrix evaluation
H. Preferred odor control system
I. Preliminary sizing criteria
J. Facility sizing and preliminary plan view drawings of the odor control system
K. Estimated cost
    1. Capital costs
    2. Operating costs

    3. Present worth cost
  L. Control philosophy
II. Dust control
  A. Significance of problem
    1. Worker health—organic dust toxic syndrome (ODTS), endotoxin
    2. Equipment protection
  B. Dust sources
    1. Mixing
    2. Composting—pile removal
    3. Curing—pile teardown
    4. Screening
    5. Traffic and housekeeping
  C. Methods of assessing dust
    1. Respirable dust—OSHA method 0600
    2. Total airborne dust—OSHA method 0500
    3. Personnel dust collection—Marple Multiple Cascade Impactor
    4. Anderson samplers—endotoxin
  D. Factors affecting dust
    1. Moisture of feedstocks
    2. Aeration
      a. Rate
      b. Direction—positive or negative
    3. Pile temperature
    4. Materials handling
    5. Traffic and housekeeping
  E. Regulatory aspects
  F. Control systems
    1. Dust collection and baghouse
    2. Hoods over screens and hoppers
    3. Misting systems
    4. Water trucks
    5. Filters in front-end loader (FEL) cabs
  G. Preliminary sizing criteria
  H. Preliminary plan view of equipment system locations
  I. Estimated cost
    1. Capital costs
    2. Operating costs
    3. Present worth cost
  J. Control philosophy

Example 3: Outline for TM on Compost Aeration—Methods and Controls

I. Evaluation of alternatives for aeration methods
  A. Composting phase
    1. Aeration requirements
      a. Biological requirements
      b. Heat removal

        c. Moisture control/removal
    2. Airflow direction and duration
        a. Negative aeration
            i. Periodic
            ii. Continuous
        b. Positive aeration
            i. Periodic
            ii. Continuous
        c. Combination
            i. Blower selection
            ii. Variable speed motors
            iii. Single speed
    3. Aeration design
        a. Above-grade
            i. Reusable piping
            ii. Disposable piping
        b. Below-grade piping
        c. Subsurface pipe and orifice
            i. Examples
                LRI, Washington
                Borland, Sweden
            ii. Design aspects
                Slope and drainage
                Cleanout system
                Airflow characteristics
                Cost
        d. Trench
            i. Examples
                Arecibo, Puerto Rico
                Davenport, Iowa
            ii. Design aspects
                Slope and drainage
                Clean-out system
                Airflow characteristics
                Cost
        e. Perforated block
            i. Design aspects
                Slope and drainage
                Clean-out system
                Airflow characteristics
                Cost
B. Curing phase
    1. Aerated curing with blowers
    2. Passive aeration
    3. No aeration
C. Amendment storage piles

- D. Matrix—advantages and disadvantages of alternative aeration methods
- E. Recommended selection

II. Materials of construction
- A. Blowers
- B. Below-grade piping
- C. Above-grade piping
- D. Valves

III. Evaluation of alternative control methods
- A. Control philosophy
- B. Maintaining aerobic conditions
- C. Meeting regulations
- D. Achieving stability and maturity
- E. Control methods and technology for composting and curing
    1. Composting
        - i. Manual temperature and time
        - ii. Automated temperature and time
        - iii. Aeration control
        - iv. Moisture control
        - v. Oxygen evaluation method and control
    2. Curing piles
        - i. Aeration
        - ii. Temperature
        - iii. Moisture
    3. Recommended selection of control system

IV. Design aspects
- A. Preliminary sizing criteria
- B. Aeration system sizing
- C. Control system specifications
- D. Preliminary process and instrumentation diagram

V. Environmental aspects and management systems
- A. Data management and record keeping
- B. Reporting and regulatory contacts

Workshops

Workshops are generally designed to provide information and stimulate discussions on the TMs, as well as other aspects of the proposed facility and its operation. Most often, they are conducted with various parties, including regulators, public officials, staff, and the public. The workshops may be conducted at different levels depending on the audience. For example, information to the staff could be very different from the information imparted to the public or regulators. The staff may require details on equipment, odor control technology, sampling and analysis, structures and their design, and other details. The regulators and the public would be more interested in the reason for selecting a particular technology or system, how odor is going to be controlled, reporting requirements, responsibilities, etc.

Design Basis Report

The design basis report will vary with the system selected and the extent of design as provided in the TMs. It basically summarizes the decisions based on the TMs and workshops. It is a guidance tool for the engineers and agency or client staff. A preliminary design of the facility layout needs to be provided. The following is an example of what a design basis report may provide. The process selected in this example is the aerated static pile (ASP). Since the ASP system may be either open, partially enclosed, or totally enclosed, it is impossible to provide here all the necessary data for a design report. The objective of this example is to provide an overview of the type of data that may be included in the design report. The design report must be detailed for the specific facility to be built.

I. Project description
  A. Site description
  B. Site design
  C. Site water management
  D. Process flow
  E. Feedstocks
  F. Amendments
  G. Material balance
    1. Quantities—dry tonnes or tons per day; projected wet tonnes or tons per day
    2. Quality—average total solids, bulk density, volatile solids
II. The precomposting process
  A. Storage requirements
    1. Delivery schedule
    2. Storage facility—hoppers, bunkers, other
  B. Material movement—conveyors, FEL, trucks
  C. Mixing
    1. Type of mixers
    2. Number
III. The composting process
  A. Basic data
    1. Residence time
    2. Aeration rate and kind (negative, positive, both)
    3. Daily composting volume
    4. Pile height
  B. Aeration system
    1. Floor system
    2. Valves
    3. Controls
    4. Fans—process number, airflow L/min/m^3 (cfm/CY), and fan pressure
    5. Piping
  C. Material movement equipment—FEL

IV. The curing process
   A. Basic data
      1. Daily volume
      2. Residence time
      3. Aeration system—air velocity
      4. Pile height
   B. Aeration floor
      1. Fans—number, airflow L/s/m^3 (cfm/CY), and fan pressure
      2. Piping
      3. Valves
      4. Controls
   C. Material movement equipment—FEL, trucks
V. Postprocessing
   A. Screening
      1. Daily volume
      2. Number of screens
      3. Material handling—conveyors, FEL, trucks
   B. Product storage
VI. Structures—Much of the information specified in the design report will depend on the structures and their design. Is the facility going to be partially or totally enclosed?
   A. Buildings sizes and type
      1. Ancillary structures—administration, worker facilities
      2. Composting and curing structures
      3. Building ventilation—air exchanges during working and nonworking hours; if enclosed airflow for each building, e.g., receiving, composting, curing, postprocessing, etc. Exhaust fans—type, number, and airflow depending on location. These could differ for the composting, curing areas, and other locations.
      4. Electrical—lighting, other
VII. Biofilter(s) or odor control—The odor control systems selected for this facility are biofilters. A biofilter would be needed for the receiving or preprocessing area, composting, and curing area. The biofilter may be built in sections.
   A. Basic data
      1. Total airflow in L.min (cfm)
      2. Airflow for each section if appropriate
      3. Temperature and humidity in location
      4. Temperature and humidity in compost area
   B. Design criteria
      1. Airflow to biofilter in L/s (cfm)
      2. Loading rate L/s/m^2 (cfm/sf)
      3. Empty bed residence time
      4. Maximum duct velocity
      5. Maximum backpressure of media
      6. Media composition, porosity, particle size
      7. Target temperature

8. Target moisture
9. Air distribution system—piping configuration—header and lateral system; valving; booster fan (size and type)
10. Water (moisture) system surface irrigation
11. Air pretreatment (humidification)

VIII. Estimated cost
   A. Capital costs
   B. O&M cost

MATERIAL, ENERGY, AND WATER BALANCES

Material Balance

Composting is primarily materials handling and aeration requirements. The calculation of the material balance needs to be done in the design phase or if the feedstocks change. One of the most important aspects of the material balance is to provide information on materials handling and the areas needed for the various processes. Most of the material handling is on a volume basis. With ASP and windrow systems and, to a lesser extent, other technologies, considerable movement of materials is done using FELs. The bucket size is very important. Eight, ten, or more cubic yard buckets can move a considerable amount of material, and therefore be more efficient and economical. Material balance also provides information on the quantity of materials, which require refining. This will allow the designer of the facility to determine the capacity of the equipment. For example, purchasing a screen will depend on the production of the amount of product to be screened. This avoids buying an oversized or undersized screen. Another product of the material balance is to provide information on the space requirements for various phases of the operation. These would be tailored to the system and technology selected. A windrow or agitated bed operation would most likely require more space per unit of material handled than an aerated static pile, or other technologies having higher pile heights. The relation of the materials balance to land areas will be illustrated.

Table 4.1 provides some information on the characteristics of several feedstocks. In calculating the materials balance, the more exact information one inputs, the better are the results. In some cases, such as with MSW, much depends on the process flow. Removing material such as cans, plastic, and other contaminants prior to composting will result in a higher bulk density than in the case where the entire mass is treated. A drum system such as DANO and Bedminster treats the entire mass, whereas other systems remove metals, plastics, stones, and other contaminants. In a facility that was designed in Hong Kong, the waste went through a magnet, countercurrent device, and screen to remove plastic and other contaminants.

Constructing the material balance can be created using mathematical formulas (Haug, 1993) or using trial-and-error attempting to arrive at 40% total solids (TS) or 60% moisture. This is done by adjusting the ratio of bulking agent and recycling material to the feedstock. Be very wary of system sales personnel indicating that their system can handle lower total solids or higher moisture content. This can often result in higher anaerobic zones and more odors that are offensive. Generally, windrow or

TABLE 4.1
Typical Characteristics of Some Common Feedstocks

Feedstock[a]	Moisture Content %	Bulk Density kg/m³ [b] (lb/CY)	Volatile Solids %
Biosolids	70–82	830–949 (1,400–1,600)	70–80
Sewage sludge	72–84	637–1,038 (1,074–1,750)	70–80
Wood chips	65	178–246 (300–415)	95
Yard waste	56	178–313 (300- 528)	95
MSW	30–50	178–477 (300–804)	77
Food waste	80	706–860 (1,191–1,450)	75
Horse manure	59–79	563–712 (949–1,200)	97
Straw	4–27	36-101 (60–170)	95
Corn straw	4–27	196 (331)	95

[a] There is considerable variation in the characteristics of feedstocks.
[b] To convert from kg/m^3 to lb/ft^3 multiply by 1.6855.

agitated bed systems may be able to work with mixes of 38% TS or 62% moisture, since they can result in greater moisture loss and porosity within a shorter period.

Several examples of material balance will be shown. The key inputs are:

- Quantity of material (mass): This is indicated in tons or tonnes on a dry weight basis. The reason for using dry weight is that the quantity is uniform. It does not change because of drying or evaporation. From this value, the wet weight of the material to be used is calculated using percent TS or moisture content. For example, if the moisture content is 58% and the dry weight is 27 tonnes (30 tons), then the wet weight = dry weight/percent moisture/100. Thus, 27/0.58 = 46.55 tonnes of wet material.
- Total solids (TS): This quantity is the inverse of moisture content. Using this quantity, the actual weight can be calculated. If one knows the moisture content of the feedstock or any other material, it can be expressed as total solids.
- Bulk density (BD): This value is used to convert the mass to volume. Most equipment for material handling is based on volume, e.g., a front-end loader (FEL) is designed for cubic yards or cubic meters of material. To convert the mass to volume the following formula is used:

BD = kg/cubic meter or lb/CY; therefore, cubic meter (m^3) = kg/BD (CY = lb/BD)

Assuming 1 wet tonne or 1,000 kg and a BD of 296.5 kg/m^3, then 1 tonne × 1,000 (kg/tonne)/296.5 = 3.37 m^3 (similarly in our system, assuming 1 ton × 2,000 lb/ton/500 = 4 CY)

- Volatile solids (VS): The volatile solids quantity expressed in percent indicates the potential rate of decomposition. It represents the total carbon rather than volatile carbon. The relationship between VS and carbon dioxide evolution can be excellent (Bach et al., 1984). The volatile solids can be determined using a scale and muffle furnace. However, it is time-consuming. Typically, a general number is used from either known values or experience. This value is used to estimate the loss of carbon dioxide during composting.

The outputs of a material balance are:

- Mixture: This is the result of the input feedstocks and bulking agents or materials, which may include recyclable material. This is one of the most important outputs, as it determines the land area needed for composting and the volume of material that needs to be handled.
- Composting losses: This is the estimated loss of volatile solids. Its exactness is not absolute. The larger the losses, the less material is handled and the less product is produced.
- Screen feed: This is the projected material that would need to be process and refined.
- Recycled bulking agent: This is the projected amount of recycled bulking material that could be reprocessed. This depends on the screen size. The smaller the screen sizes, the more recycled bulking material there is.
- Compost: This is the estimated product output.

Several factors can affect the material balance. These are:

1. Number of operating days, e.g., five-day or seven-day operations
2. The kind and characteristics of the feedstock
3. The kind and characteristic of the bulking agents

Table 4.2 illustrates a material balance for horse waste where the relationship of bulk density to the volume that must be handled. The numbers per se are not important. This example was for horse manure to be composted in a facility in Hong Kong. It was important to have an accurate measure of the bulk density. Notice that the weights and percentage of total solids and volatile solids did not change. If the facility was designed based on a bulk density of 0.45 t/m^3 (758 lb/CY), then the infeed volume would be 44.44 m^3 (58.12 CY). The result was having 23.11 m^3 (30.23 CY) of product. However, if the bulk density was only 0.35 t/m^3 (590 lb/CY), then the infeed

TABLE 4.2
Material Balance of the Composting of 20 Tonnes (22.2 Tons) of Horse Wastes as Related to Bulk Density

Material/ Process	Volume m^3 (CY)	Wet Weight tonnes (Tons)	Dry Weight tonnes (Tons)	Total Solids %	Volatile Solids tonnes (Tons)	Volatile Solids %	Bulk Density kg/m^3 (lb/CY)
Infeed	44.8 (58.6)	20.2 (22.2)	9.1 (10)	45.0	8.6 (9.5)	95.0	449 (758)
Loss			1.4 (1.5)		0.64 (0.71)		
To cure	40.9 (53.5)	17.2 (18.9)	8.2 (9.0)	47.5	7.8 (8.6)	95	420 (708)
Loss			0.8 (0.9)				
From cure	26.4 (34.5)	12.7 (14.0)	7.4 (8.1)	58.0	6.8 (7.5)	93.0	480 (809)
Infeed	57.6 (75.3)	20.2 (22.2)	9.1 (10)	45.0	8.6 (9.5)	95.0	350 (590)
Loss			1.4 (1.5)		0.64 (0.71)		
To cure	52.1 (68.2)	17.2 (18.9)	8.2 (9.0)	47.5	7.8 (8.6)	95.0	330 (556)
Loss			0.8 (0.9)				
From cure	30.0 (39.2)	11.1 (12.2)	6.4 (7.1)	58.0	6.0 (6.6)	93.0	376 (634)
Infeed	100.9 (131.9)	20.2 (22.2)	9.1 (10)	45.0	8.6 (9.5)	95.0	200 (337)
Loss			1.4 (1.5)				
To cure	90.5 (118.4)	17.9 (18.9)	8.2 (9.0)	47.5	7.8 (8.6)	95.0	190 (320)
Loss			0.8 (0.9)				
From cure	62.1 (81.2)	12.7 (14.0)	7.4 (8.1)	58.0	6.8 (7.5)	93.0	205 (345)

volume would be 57.14 m^3 (74.74 CY). At a lower bulk density, the volume to be composted would be even greater. It is obvious how the bulk density could influence a facility design and size. The accuracy of the information may be critical in the facility design, and especially if the design is being put to bid or tender.

Table 4.3 and Table 4.4 provide a material balance in the case of an ASP system for biosolids and yard waste for 1 dry ton. In this example, some values are measured and others are estimated. Table 4.3 is the ASP system when screening is done

TABLE 4.3
Material Balance for ASP Using Biosolids and Bulking Agents for 1 Dry Ton of Biosolids Where Screening Is Done after Curing

Material	Volume m³ (CY)	Total Weight tonnes (Tons)	Dry Weight tonnes (Tons)	Volatile Solids tonnes (Tons)	Bulk Density kg/m³ (lb/CY)	Solids Content %	Volatile Solids %
Biosolids	3.823 (5.0)	3.63 (4.0)	0.9078 (1.0)	0.726 (0.8)	948.8 (1,600)	25.0	80.0
Grass 5% of biosolids	0.612 (0.8)	0.182 (0.2)	0.091 (0.1)	0.091 (0.1)	382.3 (500)	45.0	90.0
Yard waste	1.682 (2.2)	0.635 (0.7)	0.363 (0.4)	0.272 (0.3)	458.8 (600)	55.0	70.0
New wood chips	1.529 (2.0)	0.454 (0.5)	0.272 (0.3)	0.272 (0.3)	382.3 (500)	65.0	95.0
Screened recycled bulking agent	5.352 (7.0)	2.179 (2.4)	1.180 (1.3)	0.999 (1.1)	531.4 (695)	55.0	83.0
Mixture	12.310 (16.1)	7.081 (7.8)	2.814 (3.1)	2.360 (2.6)	737.8 (965)	40.0	82.0
Base (recycled BA)	0.688 (0.9)	0.272 (0.3)	0.18 (0.2)	0.091 (0.1)	531.4 (695)	55.0	65.0
Cover (unscreened)	1.759 (2.3)	0.817 (0.9)	0.454 (0.5)	0.363 (0.4)	596.4 (780)	55.0	88.6
Composting losses		3.177 (3.5)	0.272 (0.3)	0.272 (0.3)			
Unscreened compost removed for cover	1.759 (2.3)	0.817 (0.9	0.454 (0.5)	0.363 (0.4)	596.4 (780)	55.0	88.6
Aerated curing (unscreened)	10.780 (14.1)	4.993 (5.5)	2.723 (3.0)	2.179 (2.4)	596.4 (780)	55.0	79.5
Curing losses		0.182 (0.2)	0.091 (0.1)	0.091 (0.1)			
Screen feed	8.946 (11.7)	4.811 (5.3)	2.633 (2.9)	2.088 (2.3)	688.1 (900)	55.0	78.7
Recycled BA	6.040 (7.9)	2.542 (2.8)	1.362 (1.5)	1.271 (1.4)	531.4 (695)	55.0	93.0
Compost	4.282 (5.6)	2.270 (2.5)	1.271 (1.4)	00817 (0.9)	688.1 (900)	55.0	63.1

TABLE 4.3 (Continued)
Material Balance for ASP Using Biosolids and Bulking Agents for 1 Dry Ton of Biosolids Where Screening Is Done after Curing

Note: During the conversion from English units to SI units the rounding off may have resulted in some error.

Assumptions:

Consolidation factor for the mixture is 95%.

Percentage recovered by the screen:

Yard waste—50% by volume

Recycle—60% by volume

Base—95% by volume

Losses during composting—10% of the volatile solids

Losses during curing—5% of volatile solids

after curing, while in Table 4.4, screening is done before curing. When curing is done after screening, less material needs to be cured and the bulking agent is better preserved. Usually the percent solids of the feedstocks and bulking agents are measured or known. The bulk density could be measured or a general value assumed. In this example, the system to be used is an ASP. The trenches or pipes would have a base of recycled bulking agent. The cover consists of unscreened compost. The example in Table 4.3 is based on composting 1 dry ton of biosolids with yard waste that contains 5% grass. The yard waste and grass were separated in the calculations since both the moisture content and the bulk density for the grass were different from those of the yard waste. The purpose here was to illustrate that sometimes even though a single waste stream is delivered, some separation of components may be advisable if there is a difference between them, and they could affect the material balance.

Other considerations in this material balance was the need for a base and cover. Often a base is not used if the aeration system does not require it. In evaluating the potential operation, it was assumed that when a pile is torn down, a portion of the material is used as an insulation cover over the new mixture to be composted. In the case of a windrow or agitated bed system, a base and a cover would not be used. The same may apply to tunnel systems.

Using the data in Table 4.4, it is possible to estimate the land area requirements for different components. This is shown in Table 4.6. Table 4.5 illustrates a material balance for a mixed waste system. The numbers of wastes to be processed are more numerous than in homogeneous waste cases such as biosolids or manure.

Energy Balance

During the decomposition process, energy is produced in the form of heat. This energy becomes very important, as the increase in temperature is needed for

TABLE 4.4
Material Balance for 1 Dry Ton of Biosolids When Screening Is Done before Curing

Material	Volume m³ (CY)	Total Weight tonnes (Tons)	Dry Weight tonnes (Tons)	Volatile Solids tonnes (Tons)	Bulk Density kg/m³ (lb/CY)	Solids Content %	Volatile Solids %
Biosolids	3.8 (5.0)	3.6 (4.0)	0.91 (1.0)	0.73 (0.8)	948.8 (1,600)	25.0	80.0
Grass 5% of biosolids	0.6 (0.8)	0.18 (0.2)	0.09 (0.1)	0.09 (0.1)	296.5 (500)	45.0	90.0
Yard waste	1.7 (2.2)	0.64 (0.7)	0.36 (0.4)	0.27 (0.3)	355.8 (600)	55.0	70.0
New wood waste	1.5 (2.0)	0.45 (0.5)	0.27 (0.3)	0.27 (0.3)	296.5 (500)	65.0	95.0
Recycled bulking agent	5.4 (7.0)	2.18 (2.4)	1.18 (1.3)	1.09 (1.2)	412.1 (695)	55.0	93.0
Mixture	12.3 (16.1)	7.08 (7.8)	2.81 (3.1)	2.45 (2.7)	573.4 (967)	40.1	86.6
Base (recycled BA)	0.7 (0.9)	0.27 (0.3)	0.18 (0.2)	0.18 (0.2)	412.1 (695)	55.0	93.0
Cover (unscreened)	1.4 (1.8)	0.64 (0.7)	0.36 (0.4)	0.27 (0.3)	462.5 (780)	55.0	88.6
Composting losses		5.26 (5.8)	0.27 (0.3)	0.27 (0.3)			
Cover (unscreened)	1.4 (1.8)	0.64 (0.7)	0.36 (0.4)	0.27 (0.3)	462.5 (780)	55.0	88.6
Screen feed	10.7 (14.0)	5.0 (5.5)	2.72 (3.0)	2.36 (2.6)	462.5 (780)	55.0	85.8
Recycled BA	6.6 (8.6)	2.72 (3.0)	1.45 (1.6)	1.36 (1.5)	412.1 (695)	55.0	93.0
Curing	4.8 (6.3)	2.54 (2.8)	1.45 (1.6)	1.36 (1.5)	533.7 (900)	55.0	92.7
Curing losses		0.09 (0.1)	0.09 (0.1)	0.09 (0.1)			
Compost	4.6 (6.0)	2.45 (2.7)	1.36 (1.5)	1.27 (1.4)	533.7 (900)	55.0	92.3

TABLE 4.4 (Continued)
Material Balance for 1 Dry Ton of Biosolids When Screening Is Done before Curing

Note: During the conversion from English units to SI units the rounding off may have resulted in some error.

Assumptions:

Consolidation factor for the mixture is 95%.

Percentage recovered by the screen:

Yard waste—50% by volume
Wood waste—70% by volume
Recycle—50% by volume
Base—95% by volume
Losses during composting—10% of the volatile solids
Losses during curing—5% of the volatile solids

pathogen and weed destruction. Excessive heat, if contained, can reach combustion levels. The elevation in temperature will result in drying, and therefore will require the addition of water, especially for certain systems. Haug (1993) discusses energy, especially the mathematical approaches to the energy balance, in detail. This aspect will not be repeated. Temperature control and management are important, since a major objective in the composting process is to produce a hygienic product. It is easier to control the temperature in static systems, depending on blowers for aeration. Most of these systems will have temperature control. In turned systems, it is more difficult to control the temperature. Regulations for pathogen and vector attraction are based on time-temperature.

The higher the volatile solids or available carbon, the higher the energy potential. MSW usually has low available carbon since paper can be a major constituent. The C:N ratio is high, and therefore a source of nitrogen may be needed to accelerate the composting process. This is often done using biosolids. Food waste, grass, and biosolids have a low C:N ratio as a result of the high nitrogen content. A woody bulking agent is therefore used to provide carbon and avoid nitrogen losses in the form of ammonia.

WATER BALANCE

Generally, the water balance is not calculated. Water balance consists of the feed water. This includes the moisture content of the feedstock; bulking agent, both new and recycled; added water; water produced during decomposition; and the final water content of the product. The three most important water contents are the feed material, added water, and product water. From a design viewpoint, the added water is important, since it may require a source of water.

Knowledge of the feed water is very important, since as was indicated earlier, the mix moisture content needs to be near 60% (total solids 40%). Added water is

TABLE 4.5
Material Balance for City Considering Mixed Waste

Material	Volume m³ (CY)	Total Weight tonnes (Tons)	Dry Weight tonnes (Tons)	Volatile Solids tonnes (Tons)	Bulk Density kg/m³ (lb/CY)	Solids Content %	Volatile Solids %
Manure	170.4 (222.9)	48.0 (52.9)	27.78 (30.6)	26.96 (29.7)	281.7 (475)	57.8	97.0
Grocery waste	130.36 (170.5)	51.5 (56.7)	3.54 (3.9)	3.64 (3.9)	394.3 (665)	6.9	99.0
MSW	468.70 (613.0)	51.5 (56.7)	18.52 (20.4)	14.25 (15.7)	109.7 (185)	35.9	77.0
Yard waste processed	30.10 (45.9)	14.9 (16.4)	8.17 (9.0)	6.54 (7.2)	211.7 (357)	55.0	80.0
New wood chips	59.10 (77.3)	9.4 (10.4)	7.26 (8.0)	6.54 (7.2)	80.1 (135)	76.7	90.0
Screened recycled BA	276.79 (362.0)	74.0 (81.5)	40.67 (44.8)	24.42 (26.9)	266.9 (450)	55.0	60.0
Mixture	1120.00 (1464.7)	249.3 (274.6)	105.94 (116.7)	81.16 (90.5)	222.3 (375)	42.5	77.6
Composting loss			10.62 (11.7)	10.62 (11.7)			
Screen feed	689.82 (902.2)	184.3 (203.0)	95.32 (105.0)	71.53 (78.8)	266.9 (450)	55.0	74.3
Screened recycled BA	276.8 (362.0)	74.0 (81.5)	40.67 (44.8)	24.42 (26.9)	266.9 (450)	55.0	60.0
Aerated curing	412.88 (540.0)	99.4 (109.5)	54.65 (60.2)	50.11 (55.2)	240.8 (406)	55.0	93.0
Curing loss		4.5 (5.0)	2.54 (2.8)	2.54 (2.8)			
Compost to storage	356.13 (469.7)	84.8 (93.4)	52.20 (57.5)	47.57 (52.4)	240.8 (406)	55.0	91.3

Note: During the conversion from English units to SI units the rounding off may have resulted in some error.

Assumptions:

Losses during composting—10%

Losses during curing—5%

needed to maintain a moisture content ranging from 40 to 55%. Turned and agitated bed systems will lose more water by evaporation than static systems, and therefore added water is needed. Water management can be effective in odor control in windrow systems. Screening and refining need to be done at a moisture content between 40 and 45%. Mathematical calculations can be found in Haug's book (1993) and are not repeated.

TABLE 4.6
Estimated Land Area Requirements for Several Components in Table 4.4 (see note below)

Component	Daily Volume (CY/m³)	Design Calendar (Days)	Value Operating (Days)	Total Volume (CY/m³)	Volume (CF/m³)	Pile Height (FT/m)	Area (SF/m²)	Side Dimension (FT/m)	Dim. w/ Slope (FT/m)	Dim. w/ Access (FT/m)	Total Area (SF/m²)
	0.765			0.765	0.028	0.305	0.093	0.305	0.305	0.305	0.093
Yard waste-EN	3	5	4	12	324	12	27	5	17	47	2,227.48
Yard waste-INT	**2.30**	**5**	**4**	**9.18**	**9.17**	**3.66**	**2.51**	**1.58**	**5.24**	**14.39**	**207.16**
New wood chips-EN	2	10	7	14	386	12	32	6	18	33	1,089.00
New wood chips-INT	**1.53**	**10**	**7**	**10.93**	**10.92**	**3.66**	**2.99**	**1.73**	**5.39**	**10.07**	**101.28**
Recycle-EN	8	5	4	32	864	12	72	8.5	20.5	50.5	2,548.76
Recycle-INT	**6.12**	**5**	**4**	**24.48**	**24.45**	**3.66**	**6.70**	**2.59**	**6.25**	**15.40**	**237.04**
Composting-EN	16.0	21	15	240.0	6,480.0	8.0	810.0	28.5	38.5	38.5	1,479.21
Composting-INT	**12.24**	**21**	**15**	**183.60**	**183.38**	**2.44**	**75.33**	**8.68**	**11.73**	**11.73**	**137.57**
Curing-EN	14	30	21	294	7,938	10	794	28	38	53	2,827.52
Curing-INT	**10.71**	**30**	**21**	**224.91**	**224.65**	**3.05**	**73.82**	**8.59**	**11.64**	**16.22**	**262.96**
Compost storage-EN	6	90	64	384	10,368	10	1,037	32	42	72	5,212.75
Compost storage-INT	**4.59**	**90**	**64**	**293.76**	**293.41**	**3.05**	**96.42**	**9.82**	**12.87**	**22.02**	**484.79**
Total-EN											16,330.72
Total-INT											**1,518.76**

Notes:
1. EN= English; INT=international
2. Additional land will be required for: Bulking agent hoppers (one each for new woodchips and recycle); sludge receiving hoppers; mixing building; aeration station for composting and drying; screening; yard waste receiving; yard waste grinding; yard waste composting; site access and roadways

CONCLUSION

The approaches to design require communication between the engineer and the client. This is often accomplished through technical memoranda and workshops or direct discussions with the client.

Prior to any design or even planning, a materials balance needs to be done. This material balance takes into consideration inputs to the facility, volatile solids loss, and volume of materials that need to be handled. Based on the material balance, land areas for various phases of the process, e.g., composting, curing, and storage, can be determined. It is used to estimate the equipment sizes for materials handling. The material balance also provides an estimate of the volume of the product to be produced.

Energy and water balances are usually not conducted. Water addition depends on the process and system. It is added empirically based on measurements or observation.

REFERENCES

Bach, P. D., Shoda, D. M., and Kubota, H. 1984. Rate of composting of dewatered sewage sludge in continuously mixed isothermal reactor. *J. Fermentation Technol.* 62:285–92.

Haug, R. T. 1993. *The practical handbook of compost engineering*. Boca Raton, FL: Lewis Publishers.

Rynk, R., van de Kamp, M., Willson, G. B., Singley, M. E., Richard, T. L., Kolega, J. L., Gouin, F. R., Laliberty, L., Kay, D., Murphy, D. W., Hoitink, H. A. J., and Briton, W. F. 1992. *On-farm composting handbook.* Ithaca, NY: Northeast Regional Agricultural Engineering Service.

5 Facilities Planning

INTRODUCTION

Initial planning is crucial to a well-designed and operational facility. Moreover, it can avoid problems with the public, particularly adjacent receptors. It may be valuable to involve the public early in the process. What is the facility going to process? How will the facility avoid taking in hazardous waste? What plans are there for traffic? If issues arise, such as odors or noise, who is responsible? Proper planning can avoid significant odor problems, bioaerosol issues, runoff, groundwater aspects, and other environmental issues. It can result in more efficient and economic operations.

Proper facilities planning provides opportunities for growth at a future period and can avoid the need for additional permitting requirements. Versatility is an important feature in a facilities design, and a plan that calls for multipurpose equipment is more likely to keep up with rapid changes in technology and applications than one that allows for single-purpose purchases. It provides for judicious capital investment.

Several considerations need to be made in planning for a facility. The primary ones are:

- Facility location
- Technology selection
- Environmental management, especially odor control
- Compost market potential

Although an environmental impact report may not be needed by a state regulatory agency, it may be wise to conduct one internally. This environmental impact assessment could provide directions for a sound environmental facility that would avoid future problems. It could assist in obtaining the necessary local and state permits. It could also provide direction in the design and avoid potential problems.

The facilities plan should include an assessment of available and potential markets, as this will affect type and size of equipment, storage facilities, and transportation considerations.

This chapter covers the various aspects of the facility planning process. It covers such details as:

- Site selection aspects
- Permitting considerations
- Site condition considerations
- System selection considerations

This chapter is not intended as a comprehensive guide for planning a facility. Its intention is to provide direction and considerations in planning. Numerous consulting firms are very proficient in producing a facilities plan and an operational plan.

FACILITY OWNERSHIP AND MANAGEMENT

This aspect is important for communities. There are several options (USEPA, 1994). These are community/municipal facilities, merchant facilities, privatized facilities, and contract services, as shown in Table 5.1. Each of these options has different economic considerations as well as responsibilities.

COMMUNITY INVOLVEMENT

Regardless of whether the facility is community (publicly) owned and operated or privately owned and operated, community involvement is a very important part of the planning process. It must begin early in the process. This could involve the following:

- Education as to the benefits of composting and your program. Awareness that hazardous waste will not be taken into the facility. That the objective is organic recycling, reducing waste to other waste management facilities such as landfills, and reducing greenhouse gas emissions.
- Odor panels
- Complaint response procedure
- Description of successful facilities
- Brochures or newsletter

The key is communication and is discussed in Chapter 16.

Site evaluation is one of the most important phases of planning. This involves:

1. Local and state regulations and requirements. The regulations related to composting facilities could differ depending on the type and quantity of the feedstock. Thought should be given to future potential use of materials. For example, in planning and designing a yard waste facility, should one consider the potential for accepting food waste or biowaste? In the future these materials may prove to be either regulated or an important source of income.

   Regulations may affect distance to receptors and require setbacks to residences, and commercial receptors. There may be requirements related to bodies of water.

   The U.S. Environmental Protection Agency regulates pathogens in sewage sludge and biosolids under 40CFR503. Many states use the same regulations for other wastes. Therefore, the facility plan needs to consider the design and operations as related to these regulations.

TABLE 5.1
Facility Ownership and Management Options for Communities

Facility Type	Ownership	Operator	Arrangement	Advantages	Disadvantages
Public	Community/ municipality	Municipality or community	Municipality provides equipment and operates the facility	Municipality has full control	Municipality shoulders all financial aspects, responsibility, risk
Public/private	Municipality	Private contractor	Site and equipment owned by municipality; long-term contract for operation based on bids; a separate contract could be had for product marketing	Municipality has some control of operations; shared risk depending on contract arrangements	Municipality is responsible for siting and capital costs
Privatized	Private vendor	Private vendor	A private vendor is responsible for the design, ownership, and operation of the facility; a contract for tipping fees under long-term arrangement	Municipality as a result of the long-term contract minimizes the investment risk of the private vendor	Municipality does not have full control over operations
Merchant	Private vendor	Private vendor	Facility is owned and operated by vendor; vendor responsible for obtaining feedstocks	No risk to municipality	High risk to vendor

2. Location. Some of the important considerations are proximity to neighbors and feedstock sources. Proximity to feedstocks could affect transportation costs, and traffic can have an effect on neighbors. At one facility the most important complaint and attempt to close or restrict the facility was the daily truck traffic. This also prevented potential expansion and economic considerations.

   Some of the other considerations in the location of the site are visibility to neighbors, buffer zones related to distances from watercourses, odor impacts, traffic flow, distance from feedstock sources, distance to markets, and site characteristics.
3. Site characteristics. Some of the most important site considerations are slope and topography, drainage, potential for surface runoff and erosion, subsurface considerations, soil characteristics, and archaeological aspects.

   A slope of 2% is often preferable, as it allows proper drainage and avoidance of ponding. Ponding can lead to odors and be a breeding ground for mosquitoes. A greater slope may require grading and the expenses associated with the movement of materials, as well as prevention of runoff and erosion.

   Proper drainage is necessary and storm water and runoff need to be directed either to a pond or discharged to other facilities. Water stored in the pond could be used in the composting process or for fire protection. Small amounts of runoff could be discharged (depending on permitting requirements) onto a grassed area.

   Some of the soil characteristics that may need to be considered are soil type, percolation rates, and depth to groundwater.

   In some locations, archaeological information must be provided as a permitting requirement.
4. Environmental considerations. As was pointed out several times, odor aspects and potential impact to receptors require serious consideration. An odor assessment in the surrounding area could be very valuable. An existing odor source should be determined and its characteristics evaluated. This can be very valuable in the event the compost facility is accused of odors. In one legal case, it was shown that the odor that receptors smelled did not come from the composting facility but from an animal farm. An industry even located at some distance may be a source of odors. It may be valuable to conduct odor modeling once the facility design and system is selected. The odor model may suggest that the system needs to be changed, e.g., from windrow to aerated static pile, or location of certain operations enclosed or located in a different area on the site.

   Other considerations are noise and dust. Dust from grinding operations, screening, or traffic may need to be mitigated. Grinding and screening operations may need a misting system to mitigate dust. Certain roadways may need pavement.

   Visibility can influence human perception. Often people smell with their eyes. Steam produced during a composting operation may suggest to the public a source of chemical discharge. Often tree barriers can reduce visibility impacts. Appearance can also have a positive or negative impact by the public. Clean sites suggest good management.

5. Resource management. Resource management involves both the incoming feedstocks and bulking agents and the marketing of the compost. Proximity to availability of feedstocks and bulking agents reduces transportation costs. This will also minimize traffic and potential complaints.

   Prior to selecting a site, it is important to identify the sources of feedstocks, sources of bulking agents, and sources of potential markets. A compost market analysis may well be worthwhile.

PERMIT AND ANCILLARY REQUIREMENTS

Permitting requirements vary with local, county, and state entities. Requirements will also depend on the type and extent of the facility. If the facility is to be enclosed, permitting will also require building, fire, and other permits. Permitting may require information on endangered species, impacts to fish and wildlife, archaeological disturbances, and other environmental aspects. Often it is advisable to consult with various agencies and their staff. This could reduce the time needed to obtain the permit and avoid delays. It is very important to avoid antagonizing individuals who may have an impact on obtaining permits. Recognize the agency and its staff's idiosyncrasies and attitudes. The permitting process can be complex. The following is an example for the establishment of an enclosed facility in Southern California for biosolids composting. This involved not only the permitting requirements but also consultation with different agencies.

- California Environmental Quality Act (CEQA): This required an environmental impact report (EIR) and its subsequent public comment period.
- City requirements
  - Project information document to city council and planning commission
  - Site grading permit
  - Building permit
  - Draft integrated waste management plan nondisposal facility element
  - Draft condition use permit
  - Fire department permit
- South Coast Air Quality Management District (SCAQMD)
  - Emission data, factors, and assessment
  - Permit to construct
  - Permit to operate
- Water quality board
  - Notice of intent
  - Waste discharge requirement
- California Integrated Waste Management Board (CIWMB) (currently CalRecycle)
  - Solid waste facility permit
- State department of health services
  - Vector control consultation
- USEPA Region IX: Consultation on compliance with 40CFR503.

REGULATIONS

Regulations pertaining to permitting will vary with the state. One of the most comprehensive requirements is from CalRecycle. The information below illustrates the comprehensive aspects of the guide for law enforcement agencies (LEAs) in California.

CalRecycle staff developed this outline as a guide to lead agencies in the preparation of California Environmental Quality Act (CEQA) documentation. It is also provided to responsible agencies for their review of documentation for the construction or operation of a solid waste disposal facility requiring a full solid waste facility permit (SWFP) or a standardized SWFP, under the CalRecycle regulatory tiered permit framework. All of the information is pertinent to the processing and issuance of a full SWFP, or a standardized SWFP, and is of great benefit if discussed fully in an EIR or at an appropriate level of detail in a negative declaration (ND) or mitigated negative declaration (MND) developed for the issuance or revision of a SWFP. A negative declaration is an environmental document prepared when it has been determined that the project would not have any significant environmental impacts, and therefore preparation of an environmental impact report would not be required.

This outline is intended to assist the lead agency in the identification and consideration of issues that the lead agency might wish to address in the preparation of its environmental documents (EDs). This is not a list of issues that CalRecycle will require to be addressed in order to deem the ED adequate for CalRecycle approval purposes. The appropriate level of detail for an ED should be determined by early consultation and cooperation between the lead agency, local enforcement agency, and other responsible agencies, and is at the discretion of the lead agency.

I. General background information
   1. Project location
   2. Owner and operator of the facility (property owner if different)
   3. Name and registration number of site design engineer
   4. Need for project
   5. Service projections for the life of the facility taking into account AB 939 waste diversion mandates
   6. Existing facilities
   7. Regional map/surrounding area map
   8. Conformance to waste management plan (compliance with PRC Section 50000)
   9. Designation in general plan (compliance with PRC Section 50000.5)
   10. Initial study and environmental checklist

II. Project description
   1. Site description
      - Topographical map
      - Size of site (acres)
      - Site design, including but not limited to site/layout map, active compost area, feedstock storage areas, well locations, drainage features, and property boundaries

- Total site capacity for active compost (in cubic meters or yards)
- Average and maximum quantity of individual types of feedstock processed in tonnes per day or tons per day (green material, manure, MSW, wood chips, special waste, etc.)
- Maximum quantities of active compost, feedstocks, amendments, and additives on hand at one time (in cubic meters or cubic yards)
- Sources of individual types of waste received daily
- Expected facility life span
- Current land use
- Historic land use
- Current zoning
- Detailed environmental setting, including but not limited to climatological factors, physical setting, ground and surface water, soils, surrounding land use
- Classification of disposal site if sited on a landfill (SWFP number)
- Type of users of the site (commercial, public, private)
- Construction description (e.g., grading plan, drainage plan)
- List of approvals required by federal, state, and local agencies in order to implement project

2. Design and operations
    - Verification of compliance with USEPA, California Department of Health Services, Department of Toxic Substances Control, Air Pollution Control District or Air Quality Management District, Regional Water Control Board, CalRecycle, and state minimum standards for solid waste handling and disposal requirements
    - Method of composting and details of composting
        - Construction
        - Windrow, static, "in vessel"
        - Forced air, mechanical
        - Maximum height, length and width, spacing
        - Typical operation cycle, processing time for each phase
        - Evaporative emissions (volatiles)
        - Type of emission
        - Rate of production
    - Additives
        - Type, amount, and application method
        - Chemical (e.g., fertilizer)
        - Bulking agent
        - Microbial
    - Monitoring
        - Feedstock types
        - Type of test
        - Frequency
        - Responsible party
        - Reporting method
    - Provisions for handling unacceptable feedstock

- Process (composting material)
  - Type of test (temperature, moisture)
  - Frequency
  - Responsible party
  - Reporting method
- Leachate
  - Type of test (metals, pathogens, nitrogen)
  - Frequency
  - Responsible party
  - Reporting method
- Product
  - Type of test (metals, pathogens, nitrogen)
  - Frequency
  - Responsible party
  - Reporting method
- Water supply
  - Source, well or municipal, sufficiency
    - To piles, windrows
    - To grinders
    - To mixers
    - For fire suppression
    - For drinking supply
- Waste characterization
- Equipment
  - Number and types
  - Emissions
  - Stand-by equipment availability, number and type of equipment
- Operating days and hours (days/week, hours/day, start/stop times)
  - Describe the operating cycle of the facility, including hours waste is received, windrows are turned, product is removed
- Traffic number and type of vehicles
  - Access routes (ingress/egress)
  - Unloading
  - On-site roads
  - Public and commercial routing
  - Number and types of vehicles entering and leaving the site per day
  - Modifications required during inclement weather
  - Emissions
- Provisions for site security (fencing, gates, police or security protection)
- Fire controls
  - Nearest fire department
  - On-site
- Vector controls
- Litter controls

- Odor controls
- Dust controls
- Noise and vibration control provisions
  - Noise levels generated by the project (construction and operation)
  - Vibration levels generated by the project (construction and operation)
- Weight scales
- Product storage
  - Volume
  - Time
  - Location
  - Handling
- Leachate containment
  - Low-permeability barrier, pad liner
    - Material type
    - Availability
    - Permeability, moisture content
  - Collection and containment system
  - Recirculation plan
- Erosion controls
- Sedimentation controls, such as siltation basins and location of such controls
- Drainage facilities (run-on and runoff)
  - Drainage plan (can be included with site map)
- Method of handling special wastes (liquids, sludge, white goods)
- Method of handling incidental hazardous waste
  - Exclusion
  - Storage
  - Removal
- Number of employees and duties
- Site improvements
  - Drinking water (well, municipal, bottled)
  - Sanitary facilities
  - Communications
  - Electrical provisions
  - Office building
- Risk of upset
  - Contingency plan
  - Public health and safety
  - Employee health and safety

III. Existing environment

1. Climate
   - Average precipitation
     - Seasonal
     - Annual
   - Seasonal temperature range

  - Wind conditions (wind roses)
    - Direction
    - Velocity
  - Evaporation rate
    - Seasonal
    - Annual
2. Air
  - Baseline air quality data (attainment status)
  - Existing emissions
    - Equipment
    - Hauling vehicles
    - Other emission sources
  - Project emissions
    - Equipment
    - Hauling vehicles
    - Other emission sources
    - Dust including PM-10 data for project construction operations
  - Leachate evaporation
  - Odor
3. Surface water
  - Existing surface waters (streams, rivers, etc.)
  - Drainage courses
  - Average seasonal flows
  - Greatest anticipated 24-hour or 6-day rainfall amount
  - Beneficial uses of waters
  - Water quality analyses
  - Watershed characteristics
4. Subsurface water
  - Existing subsurface water (aquifer, aquiclude, etc.)
  - Beneficial uses of waters
  - Water quality analyses (site-specific tests)
  - Location of wells within 1 mile of site
  - Depth to groundwater (from site-specific tests)
5. Geology
  - Description of subsurface strata (in place)
  - Soils
    - Unified soil classification (CH, OH, etc.)
    - Soil texture, percent passing through #200 sieve
    - Liquid limits
    - Plasticity index
    - Permeability of soils (field samples)
  - Seismicity
    - Estimate of seismic risk to the site (faults underlying the site, distance to nearest fault, maximum probable earthquake (MPE), maximum ground acceleration (MGA) of fault, etc.)
    - Liquefaction potential

    - Differential settlement potential
    - Boring logs (include locations)
  - Mineral deposits (including gavels)
6. Land
   - Description of site surface
   - Maximum slope on the site
   - Slope stability
7. Flora
   - Description of site flora
   - Vegetation that will be permanently removed
   - Relation between vegetation and slope stability and erodability
   - Rare and endangered flora (including takes)
8. Fauna
   - Description of site fauna
   - Resident population of rodents and other potential vectors
   - Rare and endangered fauna (including takes)
9. Noise
   - Local noise ordinance criteria
   - Background noise levels at and adjacent to site
   - Location of noise receptors (residents, schools, hospitals)
10. Social
    - Growth inducement
11. Land use compatibility
    - Zoning
    - Adjacent land use
    - Distance to nearest residences
12. Plan consistency
    - General plan
    - Regional plan (CIWMP)
13. Historical/cultural
    - Archaeological sites
    - Historical sites
    - Cultural sites
14. Traffic
    - Existing traffic conditions
15. Aesthetics (compatible with specific general plan policies or view shed ordinances)

IV. Project-related impacts to the following environmental assessment areas or cumulative impacts and significant impacts remaining after mitigation
1. Climate
2. Air
3. Water
   - Surface
   - Subsurface
4. Geology
5. Land

6. Flora
7. Fauna
8. Noise
9. Social
10. Historical/cultural
11. Traffic
12. Aesthetics (compatible with specific general plan policies or view shed ordinances)

V. Alternatives (if required)
1. Review of alternative locations
2. Other alternatives (e.g., reduced project)
3. No project

VI. Executive summary
1. Summary of project and consequences
2. Impacts, mitigation measures, and alternatives (table, outline)
3. Areas of controversy
4. Resolution of issues

VII. Organizations and people consulted
1. Public response
2. Public meetings (date and time)
3. Contributors to report (names and qualifications)
4. Persons consulted

VIII. Mitigation reporting or monitoring program (table)
1. Identification of impacts
2. Identification of mitigation measures
3. Implementation schedule
4. Monitoring frequency
5. Responsible party
6. Enforcement method
7. Conflict resolution plan
8. Compliance with AB 314 and SB 749

As can be seen, this requirement is very involved and would require knowledgeable individuals, usually an engineering firm, to provide a document with this detail. Many locations would not require such an extensive document or information. The cost to provide such detailed information can be prohibitive.

SITE SELECTION

Location is extremely important. In selecting a site, there are several important considerations:

- Site existing conditions
- Availability and cost of land
- Neighbors

  - Proximity
  - Types
    - Residences
    - Commercial
    - Industrial
    - Agricultural
- Availability of feedstocks and product marketing including transportation routes
- Availability of utilities such as sewerage, water, power

Before a site is selected, it is important to evaluate the existing conditions of potential sites as carefully as possible. The availability and cost of the land are important factors and obvious. Site consideration involves the following.

Neighbors

The proximity and types of neighbors will affect site layout, technology selection, and structures. In general, regardless of the type of neighbor, they will not be tolerant of frequent odors. Agricultural neighbors are the least apt to complain, as they are used to odors from farm animals and other sources. Residents will be least tolerant. Proximity is another factor. Although odors can travel large distances, neighbors located beyond 1 mile are less apt to complain unless prevailing wind is directly in their direction. The management of odors (Chapter 10) can be very important. One agency in Northern California evaluated several sites. The site the staff thought would be best raised great opposition from nearby residents. Some felt that bioaerosols would affect schoolchildren playing outdoors a mile away. Others objected to traffic at an interchange. There were several residences within one-quarter of a mile from the proposed facility. After considerable expenses involving several public hearings, the project was abandoned.

Another factor that can influence neighbors' tolerance to a facility is transportation routes. Truck transportation of feedstocks or final product through residential areas will be upsetting. An example is truck transportation of biosolids from the Columbus, Ohio, wastewater treatment plant to the composting facility. This issue affecting neighbors could have been avoided with two other choices, a bridge over a small river or by having a pipe transport the biosolids to the facility. Considering the cost of fuel and labor, either one of these choices would have been more acceptable.

Availability of Feedstocks and Product Marketing Including Transportation Routes

With the cost of fuel and labor increasing, the proximity for obtaining feedstocks and transportation of the product to markets is an important economic consideration. Before selecting a location, it is important to lock in the availability of feedstocks. Any fees obtained for biosolids, food waste, and any other feedstocks will make the facility more cost-effective.

Transportation of the product to potential markets can affect the remuneration and value of the product. Compost is light, 534 to 593 kg per cubic meter (900 to 1,000 lb per cubic yard). Therefore, it cannot be transported very large distances. Bagging a product increases its distance to market. A good market for bulk products needs to be located within 40 to 80 km (25 to 50 miles) of a facility.

Availability of Utilities Such as Sewerage, Water, and Power

Availability of utilities can affect the capital cost of a facility. Sewerage is the least cost, as it primarily affects the labor force on the facility. The installation of a septic system can easily handle labor-produced waste. Water and power brought into the facility can be expensive. Power needs are a function of the technology. Windrow systems use little electric power, but on the other hand, they use gasoline, diesel, or both. Importing this fuel is also a cost factor.

SITE EXISTING CONDITIONS

Once a site is located, several other considerations need to be taken into account prior to design. These include:

- Site layout
- Traffic
- Water management and resources
- Geotechnical
- Utilities

Site Layout

After the evaluation, a preliminary site layout can be provided. This is where the material balance is crucial (see Chapter 4). The site layout may require several reiterations, which involve traffic patterns, building layouts, water management and water resources, grading if needed, power resources, fire protection, and area requirements for specific functions. The final area requirements will depend on the system selected. However, the initial site layout can be determined for such areas as administration, weight scale, feedstock acceptance and preparation, bulking agent storage, traffic patterns, and product processing and storage. Since different systems require different area configurations and size, the final site layout needs to wait until the system is selected.

Traffic

Two traffic considerations need to taken into account in the planning of a facility. The first is with regards to incoming traffic conditions. This involves both bringing in feedstocks and the transportation of product from the facility. Both traffic conditions need to consider the neighborhood. Residential areas need to be avoided. Highway egress and congestion must be taken into consideration. These conditions will avoid antagonizing neighbors and provide ease in transporting material to and from the facility.

Within the facility attention must be given to access to the weighing station for both incoming and outgoing material. In case of fire, there must be access for fire equipment. In some cases, local regulations may dictate fire lanes.

Water Management and Resources

Plans must be made for water for the composting process, dust protection, runoff, storm water management, and fire prevention. Water needs for the composting process are a function of the system to be selected. The use of a water spray while turning windrows reduces odor, emissions, and dust. It also minimizes visible vapor conditions of steam, which may appear to neighbors as an undesirable emission. Water may be needed for dust prevention on the roads approaching the facility as well as within the facility. This water may be available from pond water collected from storm water. Runoff and storm water should be collected. This water could be available for processing and fire prevention. Runoff and storm water management is also important to avoid contamination of nearby water bodies.

The U.S. Department of Agriculture, National Resource Conservation Service recommends a minimum of 91 m (300 feet) from a well, spring, or body of water. States and some local authorities may have different requirements, and these need to be considered in the planning of a composting facility.

Geotechnical

Geotechnical conditions are very site specific. However, there are some general considerations. The U.S. Department of Agriculture, National Resource Conservation Service suggests the site limits indicated in Table 5.2.

State or local authorities may have other requirements. Laws and regulations must comply with federal, state, and local laws.

TABLE 5.2
Suggested Geotechnical Conditions

Feature	Conditions
Maximum slope	8%
Maximum permeability (least permeable horizon over 12 in. thick)	5 cm/h (2.0 in./h)
Minimum depth to bedrock	76 cm (30 in.)
Minimum depth to high water table 45 cm (18 in.)	
Minimum flooding event	1.0 per 25 years
Maximum fraction of 3 in. rock (percent by weight)	35%

Utilities

The provision of utilities is an important part of design. In addition to electric power needed for mechanical equipment, odor control systems, lighting, pumps, and other aspects, fuel storage must be provided. Sewerage must be provided for employees. This may require connection to municipal facilities or having a septic system with an approved design as per state or local conditions.

SYSTEM AND TECHNOLOGY SELECTION

In Chapter 6 various technologies and systems are indicated. Some systems are nonproprietary, whereas others are proprietary. In addition to the systems indicated in Chapter 6, the brochure by Chiumenti et al. (2005) can provide additional information. Footprints and areas vary with different systems and technologies. With the nonproprietary systems, the ASP requires less space than most windrows since pile height can reach 3 to 4 m (10 to 12 feet). Most windrows have a pile height of 0.9 to 2.7 m (3 to 9 feet). Straddle type machines generally produce windrows of 1.8 to 2.1 m (6 to 7 feet), whereas the elevated conveyor type machines or front-end loaders may produce higher windrows.

Proprietary systems vary. Drum type systems require not only space for drums but also additional space to complete the composting and curing. Agitated bed systems may have piles with heights of 3 m (10 feet). Vertical systems (which there are very few of) take less space than horizontal systems.

Chiumenti et al. (2005) provide excellent illustrations of various systems. Rynk et al. (1992) provide area requirements for piles and windrows. Other sources are the Composting Council of Canada (1995) and the Composting Association of the UK (2004).

Thus, the selection of a particular system or technology will depend on several factors:

- Cost and economics.
- Location, especially with reference to residences or commercial enterprises.
- Governmental restrictions. The South Coast Air Quality District in California restricts all composting facilities to be enclosed.
- Climatic conditions. This particularly affects product storage and marketing. In cold humid climates, the market may be restricted from early spring to fall. Therefore, sufficient storage must be available for the product during late fall and winter. In warm climates where there are no limitations to product, marketing less storage space may be needed.
- Power availability. ASP or other aerated systems require blowers and hence electricity. Windrow systems require gasoline and diesel. ASP or agitated bed may not be the preferred choice if electric power is limited. A generator may be used if the power requirement is low. However, systems such as drum or agitated bed may require significant power, so that generation of electricity is not cost-effective.

CONCLUSION

Proper facility planning is extremely important for a cost-effective, well-managed operation. The planning of a facility will also affect capital costs, which will include structures, site preparation, equipment selection, and environmental considerations.

It is important in planning a facility to consider location with respect to residences or commercial enterprises, which could be impacted by traffic, odor noise, aesthetics, and visibility.

Although most people approve and favor composting as an ecological and preferred method of waste disposal, the history of composting facilities, particularly with respect to odors, has soured the attitude for their location. It is therefore important in facility planning not only to locate a site that can be easily permitted and have minimal environmental impacts, but also to communicate to residents located near the facility and local officials what the plans consist of and how the facility will be operated. Community involvement is important in the planning process.

REFERENCES

CalRecycle (California Integrated Waste Management Board [CIWMB]). http://www.calrecycle.ca.gov. Local enforcement agencies, composting facilities outline. Sacramento, CA

Chiumenti, A., Chiumenti, R., Diaz, L., Savage, G. M., Eggerth, L. L., and Goldstein, N. 2005. *Modern composting technologies*. Emmaus, PA: BioCycle.

The Composting Association of the UK. 2004. *A guide to in-vessel composting*. Northamptonshire, NNB IDH England: The Composting Association.

The Composting Council of Canada. 1995. *Composting technologies and practices*. Ottawa, Canada: The Composting Council of Canada.

Rynk, R., van de Kamp, M., Willson, G. B., Singley, M. E., Richard, T. L., Kolega, J. L., Gouin, F. R., Laliberty, L., Kay, D., Murphy, D. W., Hoitink, H. A. J., and Briton, W. F. 1992. *On-farm composting handbook*. Ithaca, NY: Northeast Regional Agricultural Engineering Service.

USEPA. 1994. *Composting yard trimmings and municipal solid waste*. USEPA 530-R-94-003. Washington, DC: Office of Solid Waste and Emergency Response, U.S. Environmental Protection Agency.

6 Composting Technologies and Systems

INTRODUCTION

The objective of this chapter is to provide the reader with an overview of the most current composting systems. Various systems will be discussed without passing judgments on the merits of one system vs. another.

The selection of a particular system or technology by a community or company may depend on several considerations. These could include:

- Economics
- Type and quantity of waste
- Potential location for a facility
- Political and regulatory aspects
- Environmental considerations
- Product quality as related to marketing

Economics not only include pricing, but also the impact of labor vs. capital costs. The economics will be discussed in detail in Chapter 8. The two items that will always increase are labor and fuel. Therefore, it is important to balance the use of equipment, which could save both labor and fuel vs. a more simple system. This is particularly true for large facilities that may need to be enclosed. For example, the use of conveyors vs. front-end loaders, mobile vs. stationary mixers, and mobile vs. stationary screens.

Many communities or agencies are forced to select the least cost system. This sometimes could be unfortunate. During the 1980s, when federal funds were available for biosolids composting, the selection of the least cost system often resulted in the selection of the poorest technology provided by the worst vendor. In order to avoid this problem, a community or agency needs to (1) prequalify vendors, and (2) write very detailed and specific specifications. In this manner, if a vendor cannot provide appropriate equipment, it cannot qualify, regardless of whether the system is the least costly.

Another consideration is ownership. There are four potential approaches:

- Publicly designed, constructed, owned, and operated
- Publicly designed, constructed, owned, and privately operated

- Privately designed, constructed, and publicly owned and operated
- Privately designed, constructed, owned, and operated

Many composting systems are applicable for small operations. A university, prison, park, or recreation center may select a totally enclosed small system, which would become uneconomical on a large scale. For example, Disney World in Orlando uses a within-vessel Wright system for its food waste. Homogeneous wastes such as biosolids, pharmaceutical, or food processing waste generally do not require the extent of preparation and removal of contaminants as would MSW.

The location of the facility will also affect not only the technology to be selected, but also enclosures and other capital investments. A facility to be located near residences or commercial enterprises needs to have greater odor and emissions control than a facility to be located in a rural setting.

Political considerations also come into play. Elected officials are often influenced by demographics. On several occasions, communities felt that they were being dumped upon as a result of having numerous undesirable facilities, such as prisons and landfills. These could impact property values, and therefore the community did not want to have a composting or similar type facility nearby. Even though they were in principle in favor of composting, they did not want it "in my back yard."

Regulations could affect facility cost and the best type of system to meet the regulations. In the United States, federal quality regulations apply only to sewage sludge, septage, and biosolids (USEPA, 2003). These regulations pertain to pathogens and control of vector attraction. Earlier regulations were on the chemical quality of sludge and biosolids (USEPA, 1994). There are no federal regulations on the physical properties of compost. Canada provinces regulate compost quality. For example, British Columbia regulates pathogens, heavy metals, and physical properties such as foreign matter. Many states in the United States adopted the federal regulations and applied them to numerous feedstocks. One of the major values of using the federal regulations in composting is that doing so provides confidence to the user that the product is safe.

Environmental considerations are very important in facility siting and the selection of an appropriate type system. Meteorological conditions, such as prevailing winds and inversions, can affect odor reception by residents nearby. The percentage occurrence of inversions needs to be considered, as these could affect surrounding areas. It is often stated that people smell with their eyes. If, because of frequent weather conditions, steam coming off piles is not dissipated, an observer may conclude that the facility is releasing undesirable emissions. Tree barriers could reduce visibility. Some microclimatological impacts can be reduced with shelterbelts (Figure 6.1). These not only reduce the visibility of a site, but also reduce dust conditions and some odors. By reducing wind velocity, there is less evaporation and drying out of windrows.

Product quality and product marketing should be an important consideration in system selection. The better the product quality, the greater its potential utilization. Refining can be expensive. A high-quality product will command a higher price per unit. Most compost products are marketed in bulk. The high cost of transportation limits the extent of distribution in bulk. A combination of bulk production, as

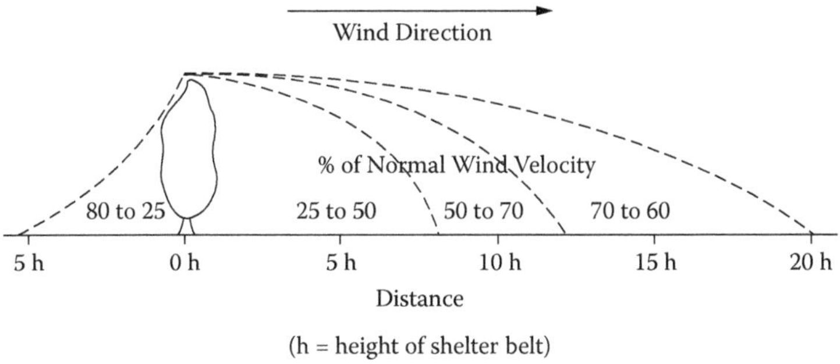

FIGURE 6.1 Effect of a shelterbelt on wind velocity.

well as bagged products, often provides the best financial arrangements. Diversity of products can also enhance sales and economics. At the Davenport, Iowa, facility sales consist of both bulk and bagged. People bringing in yard waste will often pick up bags of compost.

In the following sections, a description of the classification of composting technologies is presented along with numerous examples. It is impossible to list and describe all the available systems that exist today. The systems and facilities that I will describe are predominantly in the United States because of my extensive familiarity with them. Obviously, there are numerous others. Systems are described under a generic heading.

The reader is referred to the following sources on other facilities:

Haug, R. T., *The Practical Handbook of Compost Engineering*, Lewis Publishers, CRC Press, Boca Raton, FL, 1993
Chiumenti, A., Chiumenti, R., Diaz, L., Savage, G. M., Eggerth, L. L., and Goldstein, N., *Modern Composting Technologies*, BioCycle, Emmaus, PA, 2005
The Composting Council of Canada, *Composting Technologies and Practices*, The Composting Council of Canada, Ottawa, Ontario, Canada, 1995

During the past twenty-five years, numerous large composting facilities have been closed. Table 6.1 lists some of the more notable ones. These include systems by DANO, Purac, Buhler, Inc., Taulman-Weiss (Kneer), American Bio Tech, Bedminster, Fairfield-Hardy, and Ashbrook, as well as several large yard waste windrow facilities.

GENERIC CLASSIFICATION OF COMPOSTING TECHNOLOGIES

Haug (1993) classified composting systems according to reactor type. The following is his classification:

I. Nonreactor systems (open)
  A. Agitated solids bed (windrow)
    a. Naturally ventilated

TABLE 6.1
Closed Composting Facilities in the United States

Facility and Location	Facility and Location
Agripost, Florida	Reuter, Minnesota
Recomp, Minnesota	Reidel, Washington
Taulman Weiss, Oregon	Site II, Maryland, ASP
Cobb County, Georgia	Pembroke Pines, Florida
Pigeon Point, Delaware	Wright County, Minnesota
Long Island Composting, New York	Hartford, Connecticut
Fort Lauderdale, Florida	Plattsburg, New York
Henrico County, Virginia	Purac, Virginia

    b. Forced aeration
  B. Static solids bed
    a. Forced aeration (aerated static pile)
    b. Natural ventilation (nonaerated piles)
II. Reactor systems (in-vessel or enclosed)*
  A. Vertical solids flow
    a. Agitated solids bed
      i. Multiple hearths
      ii. Multiple floors, decks, or belts
    b. Packed bed (silo reactors)
      i. Countercurrent air: solids flow
      ii. Cross-current air: solids flow
  B. Horizontal and inclined solids flow
    a. Tumbling solids bed (rotary drums or kilns)
      i. Dispersed flow
      ii. Cells in series
      iii. Completely mixed
    b. Agitated solids bed (agitated bin or open channels)
      i. Circular shape
      ii. Rectangular shape
    c. Static solids bed (tunnel shaped)
      i. Push type
      ii. Conveyor type
  C. Nonflow (compost boxes)

* The term *in vessel* was coined by the USEPA in 1993 under Federal Regulations Part 257. This term was continued in the 40CFR503 until a revised publication in 2003. The term was then changed to *within vessel*. This was to indicate that any system could be totally enclosed and was not meant to imply a specific type of system. For example, the LRI facility in Washington state is an enclosed windrow. There are several enclosed aerated static pile facilities, such as the one in Davenport, Iowa. *Within-vessel* is a better term for describing an enclosed facility.

I have simplified the generic classification of systems and classified them into two major categories: static and turned or agitated. In some cases, a combination of aeration and turning is used. These categories can be further broken down as follows:

A. Static systems
   a. Passively aerated windrows
   b. Forced aeration—static pile
   c. Bin/container/bag/tunnel
   d. Silo/vertical reactors
B. Turned or agitated systems
   a. Windrow
   b. Drum/kiln
   c. Agitated bed

Static Systems

Passively Aerated Windrows

Most static systems provide aeration by blowers, i.e., forced aeration. However, one static system, termed passively aerated windrow, relies on convective air to provide oxygen and achieve temperatures and stabilization (Mathur et al., 1990; Rynk et al., 1992; Liao et al., 1994). The process uses perforated pipes open to the atmosphere. Feedstock with a bulking agent is piled over the pipes. As the center of the pile heats up, cool air is drawn through the pipes and provides oxygen. Figure 6.2 shows how the system is built and used.

The process has been used for animal mortalities and manure slurries in Canada. In 2002, this method was evaluated for the Oregon Department of Environmental Quality as a possible low-cost system for managing commercial food waste (Epstein et al., 2005). The results showed that this process met the 40CFR503 time and temperature requirement of 55°C or higher for three days. The vector attraction reduction requirements of the 40CFR503 regulations were met, and a stable product was produced as defined by the U.S. Compost Council's Test Methods for the Examination of Composting and Compost (TMECC) procedure. Pathogen indicator organisms (fecal coliform and *Salmonella*) were reduced to levels below the 40CFR503 regulation limits. This process very quickly achieved pathogen reduction temperatures and consistently maintained high temperatures throughout the composting period.

Passive aeration is not an approved U.S. Environmental Protection Agency (USEPA) method for pathogen reduction for the use of sewage sludge or biosolids. Some states using composting of other wastes may approve it. It has potential as a low-cost technology for use by farmers for composting animal waste.

Forced Aeration

The forced aeration system was developed in 1975 by USDA researchers at Beltsville, Maryland, and was termed the aerated static pile (ASP) method (see Figures 6.3 and 6.4). Figure 6.3 is a diagram describing the method. Figure 6.4 shows an individual ASP in use (Epstein et al., 1976). It is nonproprietary. The

FIGURE 6.2 Passive aeration pile or windrow under construction and in use.

original method was developed using negative air, i.e., suction. The objective was to reduce odors by sucking the air through pipes (negative aeration) and filtering the air into a biofilter. This was very effective in significantly reducing odors. Since then, the method includes using positive air, i.e., forcing the air through the pile. There were two advantages to the positive air system. First, forced air reduced the head loss, and therefore the energy required. Second, an external biofilter was not required. However, the potential for odors is greater. Some degree of odor control with positive aeration can be achieved using a pseudobiofilter. This is accomplished by using screened or unscreened compost over the pile. Alternatively, fabric covers can be used (see Chapter 10).

One variation to the single pile method was termed extended pile (Figure 6.3 and Figure 6.5). Daily production of feedstock was piggybacked alongside the previously formed pile. This reduced the footprint of the system and saved space.

In the early 1980s, considerable research on the ASP method was conducted at Rutgers's University, New Jersey (Singley et al. 1982). One of the major controversies

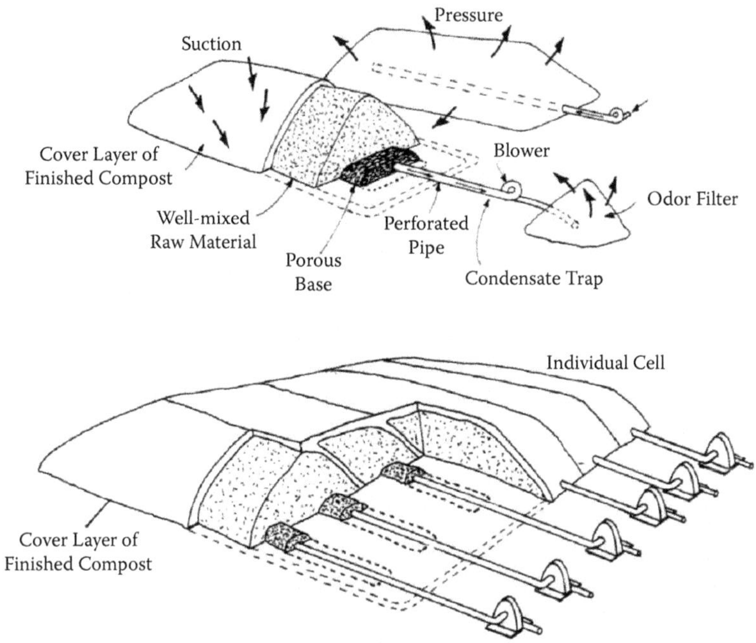

FIGURE 6.3 Upper figure shows the basic aerated pile using either negative or positive air. The lower figure shows an extended pile used to conserve space. (From Epstein et al., *J. Water Pollut. Control Fed.*, 48, 688–94, 1976; Willson et al., *Manual for Composting Sewage Sludge by the Beltsville Aerated Static Pile Method*, EPA -600/8-90-022, Municipal Environmental Research Laboratory. Cincinnati, OH, 1980.)

FIGURE 6.4 A single pile design and operation of an ASP system.

during this early research period was with regard to the temperature regime. Should the temperature regime be kept at mesophilic temperatures (25 to 45°C, 77 to 104°F) during the early phases of the process, or should the temperature reach thermophilic temperatures and be maintained for several days? Pathogen destruction needs high, thermophilic temperatures (see Chapter 12). However, the rate of organic matter decomposition is faster at lower mesophilic temperatures. Rutgers scientists proposed that initially the temperature should be kept at mesophilic temperatures to achieve greater decomposition, and this regime would follow with temperatures to destroy pathogens. The USDA team felt that, in order to achieve high thermophilic temperatures for pathogen destruction, as much energy as possible was needed. This would occur in the early stages of composting. After pathogens were destroyed, the temperature could be lowered to the mesophilic stage for greater and faster stabilization. This later procedure was adopted by USEPA in its regulations.

There are several major aspects to the ASP and numerous configurations of the system:

- Totally open.
- Partially enclosed with a roof or sides.
- Totally enclosed.
- Temperatures for pathogen destruction must occur throughout the mass that is to be composted. This is achieved through proper insulation and aeration.
- Temperatures to meet pathogen destruction must occur in three consecutive days as specified by USEPA 40CFR503 or state regulations.
- The aeration system must be designed to provide uniform and oxygen levels exceeding 10% throughout the mass.
- Sufficient porosity must occur within the pile to provide adequate aeration; this is usually achieved by varying the bulking materials ratio.

Small facilities use front-end loaders and portable mixers to combine the feedstock with bulking agents. Aeration is typically provided with disposable perforated pipes. Larger facilities could use hoppers, conveyors, pug mills, or other stationary mixers. Aeration is provided using trenches, spigots, or other systems. Details of this equipment are provided in Chapter 7.

Figure 6.5 shows the ASP system in an extended pile mode. Notice the disposable pipes used for aeration. The blowers along the concrete wall can be used for either positive aeration or negative aeration with the air sent to a biofilter.

As indicated earlier, it may be necessary to either partially enclose an ASP system (Figure 6.6) in humid areas or totally enclose a facility as a result of location, rainfall, and concern for odors (Figure 6.7).

There are several enclosed or within-vessel ASP composting plants. In these, aeration is built into the floor and provided by either trenches or spigots. The feedstock and bulking agents are placed in hoppers. Material is conveyed to mixers and then delivered to the aeration system by a combination of conveyors and front-end loaders. The discharged air is passed through a biofilter to remove odors and volatile organic compounds. An example is the facility in Davenport, Iowa (Figure 6.7). The inside of the ASP system is shown in Figure 6.8.

FIGURE 6.5 The ASP system shown in an extended pile mode.

FIGURE 6.6 A covered ASP composting facility in Harrisburg-Rockingham, Virginia.

FIGURE 6.7 An enclosed ASP facility in Davenport, Iowa. The large pipe along the wall is the air handling system for the biofilter.

FIGURE 6.8 The inside of the Davenport, Iowa, ASP composting facility. The composting piles are over a trench aeration system.

Another example is the largest facility in North America for biosolids located near Chino, California, which was designed after the Davenport, Iowa, concept. The facility began operations in April 2007 and reached full capacity in December 2008. At full capacity, the Inland Empire Regional Composting Facility (IERCF) processes 272,340 tonnes (300,000 tons) per year of recycled materials and produces approximately 183,504 cubic meters (240,000 cubic yards) of compost. The IERCF produces compost using the aerated static pile (ASP) composting method by mixing together biosolids and other organic material in large piles instead of in traditional windrows. Aeration occurs as air is pushed or pulled through the ASP. IERCF pulls air directly from the piles and sends the air to a biofilter along with air exchanged through the composting building. The biofilter is 38,230 cubic meters (50,000 cubic yards) of a special blend of wood chips that treats all of the air by removing odors and other regulated compounds before it is exhausted to the atmosphere.

The entire composting process at the IERCF takes approximately sixty days. Active composting lasts approximately twenty-two days before the pile is screened and moved into curing. The material will stabilize in curing for between thirty and thirty-eight days. After curing, the compost is ready for distribution and use. Larger materials screened from the compost are recycled back to the beginning of the composting process to be used again. Figure 6.9 shows an aerial view of the facility.

The bay design in Comox, British Columbia, is an illustration of an enclosed ASP built in units (Figure 6.10). Each unit contains one week of material (Figure 6.11). After four bays are completed, the first bay is emptied and then refilled. Obnoxious air is removed from above the pile and sent to a biofilter. Engineered Compost System (ECS) of Washington state provides a similar system, as shown in Figures 6.12 and 6.13.

FIGURE 6.9 Aerial view of the largest biosolids composting facility in the United States. (Courtesy of Jeff Ziegenbein, IEUA, Chino, California.)

FIGURE 6.10 A bay or bin system using the ASP method in Comox, British Columbia.

FIGURE 6.11 The material to be composted over an aerated trench system in the bin in Comox, British Columbia.

FIGURE 6.12 Aerated bins in Granby, Canada. (Courtesy of ECS, Seattle, Washington.)

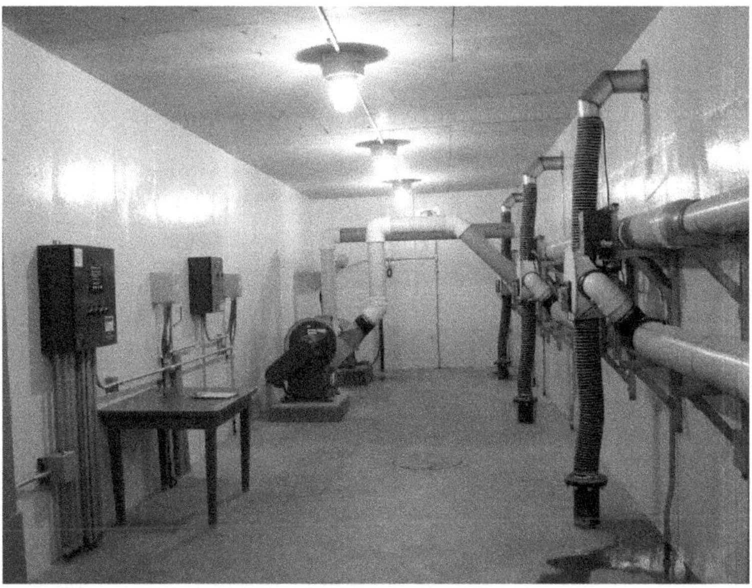

FIGURE 6.13 Fan room for the aerated bins in Granby, Canada. (Courtesy of ECS, Seattle, Washington.)

Bin/Container/Bag/Tunnel

These systems principally apply to small facilities and can be very effective in odor control. The systems are usually ventilated and are horizontal. The main differences between the various systems are the way these are loaded, unloaded, and ventilated. Loading can be done in several different ways. They can be loaded or unloaded with front-end loaders. Some European systems use telescoping conveyors to load. Others use specialized equipment to load. Unloading is done using front-end loaders, moving floors, or the bin/container is tipped to slide out the material. An example of some container types is the Wright system in Disney World, Orlando, Florida (Figure 6.14) and Engineered Compost System (ECS) of Seattle, Washington (Figures 6.15 and 6.16). There are numerous other similar systems, such as NaturTech and Green Mountain Technologies, which use modified roll-off containers.

An example of the tunnel system was built in Hamilton, Ohio, in the 1980s; it used a push plate to move the material through the tunnel and discharge it through the back end. Another example is the GICOM tunnel system in Maine. Blowers with an aeration system in the floor usually provide aeration for these systems. The air is discharged into biofilters.

An example of the bag system is the EcoPOD system by Ag-Bag that is used by Norcal, a waste management company north of San Francisco. A blower connected to perforated pipes (Figure 6.17) supplies air.

Most bin, container, and bag systems are used for relatively low volumes of feedstocks and where the location can be sensitive to odors. The systems are generally

FIGURE 6.14 Wright container system in Disney World, Orlando, Florida.

FIGURE 6.15 Composting bins being unloaded. (Courtesy of ECS, Seattle, Washington.)

proprietary. These systems will require a mixing and final preparation of the product through screening or other techniques. These systems may require space for curing, although the same container, bin, or bag could also be used for curing. They require aeration and odor control.

Essentially, no new tunnel systems have been built in the United States since the 1980s.

Silos and Vertical Reactors

Silos and vertical reactors were common in the United States during the 1980s. Today they are not being built, and many have been discontinued. Figure 6.18 shows one of the first, which was built in Portland, Oregon. This was a Taulman-Weiss system, which later became quite common. Kneer systems were built in Disney World, Orlando, Florida; Lancaster, Pennsylvania; Clayton County, Georgia; and other locations. Other vertical systems designed as rectangular buildings were built by Purac in Henrico County, Virginia; in Hartford, Connecticut; and Schenectady, New York, by American Biotech. The principal problems were excessive compaction, poor aeration, and difficulty in extracting the material. The one in Hartford, Connecticut, caught on fire and was destroyed. To overcome these problems, some European and Asian systems designed vertical systems with multiple floors and, using an agitated system, moved the material from one floor to the next. The advantage to these systems was space. The EBARA system in Sapporo, Japan, is such an example.

FIGURE 6.16 Aeration system for portable composting bins. (Courtesy of ECS, Seattle, Washington.)

Turned or Agitated Systems

Windrow

The windrow system is nonproprietary and by far has the greatest number of facilities in the United States. There are over four thousand facilities, predominantly using yard waste. These are essentially operated outdoors. The equipment used for turning windrows is proprietary and is offered by numerous vendors. Figure 6.19 shows the most commonly used turning system where the machine straddles the windrow and agitates the material. Under this condition, emissions are great, and odors can be a significant problem. A major advantage to this system is the large volume of material it can handle. It also does an excellent job of mixing and pulverizing the material. Another windrow system lifts the windrow and deposits the material along its side (Figure 6.20). In this picture the unit is

FIGURE 6.17 The Ag-Bag system and loading machine. Aeration provided by blowers.

illustrated in the LRI facility in the state of Washington. There are also several outdoor facilities.

The turning machine cuts into the windrow and stacks the material against the face of a formed windrow. This produces a continuous batch. The enclosed LRI facility in Washington uses that system but also provides aeration through spigots in the floor. This concept minimizes the potential for anaerobic conditions.

FIGURE 6.18 Vertical silo system by Taulman-Weiss in Portland, Oregon.

FIGURE 6.19 A windrow turning machine straddling the windrow.

FIGURE 6.20 Windrow turning machine, which lifts the windrow and deposits it along its side. In this case, the unit is used in the LRI enclosed facility.

The windrow system used for composting biosolids is regulated by USEPA 40CFR503. This regulation requires that windrows achieve 55°C (131°F) for fifteen consecutive days with a minimum of five turnings of the windrow. The objective is to subject the entire mass to high temperatures for pathogen destruction.

Windrows vary in width and height depending on the equipment. Generally, most windrows are 1.5 to 2.7 m (5 to 9 feet) high and 2.7 to 6.1 m (9 to 20 feet) wide with space in between for the turning machine. The turning machine straddles the windrow and mixes the material so that, over the period of the composting, all the particles will meet the temperature requirements. Aeration is provided primarily through convective airflow. As the windrow heats in the center, air is drawn from the sides. Turning not only provides mixing, but also improves porosity and breaks up the particles. Windrow systems can turn anaerobic, since within a short period after turning oxygen levels decrease. In Europe in one facility, I observed that the windrow was over an aeration system. Air was withdrawn by suction and sent to a biofilter. This not only improved aeration but also reduced odors.

Agitated Bed

There are numerous variations of the agitated bed. These are horizontal systems using turning machines, paddles, or other turning devices. International Processing System (IPS) produces the most common agitated bed in the United States. Examples of this system can be found in Burlington County, New Jersey; Greensboro, North Carolina; and other locations. The agitated bed systems are principally used in the United States for composting biosolids. They are all enclosed. Figure 6.21 shows the agitator moving through a bay in Greensboro, North Carolina, and an agitator designed and used in IPS facilities. Figure 6.22 shows an IPS agitator.

FIGURE 6.21 An IPS agitated bed system in the bay at Greensboro, North Carolina. (Courtesy of Siemens Water Technology Corporation.)

In Figure 6.23, an agitated bed system is shown with blowers in an alley alongside the bin. These provide aeration. The agitator daily moves material, and a segment of the material is discharged at the end. This follows the required period of composting for about twenty-one days. The discharged material is deposited onto a conveyor, which moves the material to the curing area. Alternatively, material is discharged into an area beyond the bay and is picked up by a front-end loader (FEL) for placement in the curing area.

The facility in Edmonton, Canada (Figure 6.24), uses the Italian Sorain Cecchini Tecno system for composting MSW and sewage sludge. This is a turning auger system traveling on a moving bridge. This facility combines an up-front drum system followed by the agitated bed system (Figure 6.25). Following composting and either before or after curing, the compost is cleaned and undesirable material removed. This can be accomplished by screening and air classification. Screens can be either portable or stationary.

FIGURE 6.22 The IPS agitator. (Courtesy of Siemens Water Technology Corporation.)

FIGURE 6.23 Agitated bed system showing the blowers used to provide additional aeration besides turning.

FIGURE 6.24 The Edmonton, Canada, MSW and biosolids composting facility using a drum system.

FIGURE 6.25 The Edmonton, Canada, MSW and biosolids composting facility showing the drum and agitated bed system.

Drum

Drum or rotating systems have been used in many facilities in Europe, but to a very limited extent in the United States. The largest facility in North America is located in Edmonton, Canada. It uses elongated drums to mix the solid waste and biosolids (Figure 6.24). The mixture is then composted in an agitated bin system (Figure 6.25). DANO is one of the largest drum systems worldwide. A DANO system was built for MSW in Portland, Oregon, but due to odor problems and other operating issues, it was dismantled and the drums moved to South Dakota (Figure 6.26). HotRot and Rotocom use smaller drums. Two facilities in Hong Kong use drums essentially for mixing (Figure 6.27). Limited temperatures are obtained as well as limited biological degradation of the feedstock.

Retention time in the drum varies with the technology. Since composting and curing are incomplete, either an ASP or a windrow system may be needed to stabilize the compost.

Bedminster is a drum system that is used in several facilities for MSW in the United States. This system was invented by Eweson and termed a digester. Facilities were built in Big Sandy, Arizona; Siever County, Tennessee; and Nantucket and Marlboro, Massachusetts.

The drum does not provide for complete composting. Retention in drums is usually from twenty-four hours to seven days, depending on manufacturer specifications. Following that period, additional composting and curing are usually done in aerated bays or windrows.

FIGURE 6.26 The DANO drum system in Rapid City, South Dakota.

FIGURE 6.27 Animal waste composting project in Hong Kong showing drum and composting bins in background.

CRITERIA FOR SYSTEM SELECTION

Table 6.2 provides some information for criteria for system selection. There are many factors going into selecting a specific technology. Some of these are:

- Economics and cost
- Location
- Amount of material to be handled
- Type of feedstock
- State, county, or local regulations

CONCLUSION

There are numerous types of composting systems in the world. The majority of the systems are operating in North America and Europe. In North America, yard waste or biosolids are the predominant feedstocks being composted. There are over four thousand yard waste and over three hundred biosolids composting facilities in the United States. In Europe the predominant feedstocks are MSW and biowaste.

The major reason why MSW is not composted largely in the United States is that landfilling is less costly for communities. Composting of MSW is much more

TABLE 6.2
Some Criteria for System Selection

System	Volume	Location to Potential Receptors: 0.8–1.609 km (0.5–1 mile)	Location to Potential Receptors: >1.609 km (>1 mile)	Primary Feedstock Used: Yard Waste	Primary Feedstock Used: Sludge/Biosolids	Primary Feedstock Used: Food Waste	Primary Feedstock Used: MSW
Windrow	High		X	X			
Windrow with covers	High	X	X		X		
ASP individual[a]	Low to medium	X	X		X	X	
ASP extended	Medium to high	X	X		X		
Within vessel	Medium to high	X	X		X		X
ASP	Medium to high	X	X		X	X	X
Agitated bed	Medium to high	X	X		X		X
Container/bag	Low	X	X		X	X	
Bin	Low to medium	X	X		X	X	

[a] The use of covers provides good odor and emissions control, thus allowing the site location to be 0.8 km (0.5 mile) from receptors.

complex in order to produce a clean, marketable product. MSW is not a homogeneous feedstock in comparison to biosolids.

In the United States, federal regulations for the management of sewage sludge, as stated in 40CFR503, helped promote composting of sewage sludge and biosolids. Land application, generally, is the least costly method for sewage sludge management, followed by composting.

The selection of an appropriate system depends on the feedstock, location, economics, and state or local regulations.

REFERENCES

Epstein, E., Willson, G. B., Burge, W. D., Mullen, D. C., and Enkiri, N. K. 1976. A forced aeration system for composting wastewater sludge. *J. Water Pollut. Control Fed.* 48:688–94.

Epstein, E., Servo, S., Ettlin, L., Cutler, A., and Sullivan, D. 2005. Composting of commercial food waste: Pathogen and related environmental issues. *J. Residuals Sci. Technol.* 2(2):79–89.

Haug, R. T. 1993. *The practical handbook of compost engineering*. Boca Raton, FL: Lewis Publishers, CRC Press.

Liao, P. H., Vizcarrra, A. T., Chen, A., and Lo, K. V. 1994. Composting of salmon farm mortalities with passive aeration. *Compost Sci. Util.* 2(4):58–66.

Mathur, S. P., Patni, N. K., and Levesque, M. P. 1990. Static pile, passive aeration composting of manure slurries using Peat as a bulking agent. *Biol. Wastes* 34:323–33.

Rynk, R., ed. 1992. *On-farm composting handbook.* Ithaca, NY: Northeast Regional Agricultural Engineering Services.

Singley, M. E., Higgins, A. J., and Frumkin-Rosengaus, M. 1982. *Sludge composting and utilization: A design and operating manual.* New Brunswick, NJ: N.J. Agric. Expt. Sta. Cook College, Rutgers.

USEPA. 1994. *A plain English guide to the EPA Part 503 Biosolids Rule.* Washington, DC: U.S. Environmental Protection Agency, Office of Wastewater Management.

USEPA. 2003. *Control of pathogens and vector attraction in sewage sludge.* EPA/625/R-92/013. Cincinnati, OH: U.S. Environmental Protection Agency, Office of Research and Development, National Risk Management Research Laboratory, Center for Environmental Research Information.

Willson, G. B., Parr, J. F., Epstein, E., Marsh, P. B., Chaney, R. L., Colacicco, D., Burge, W. D., Sikora, L. J., Tester, C. F., and Hornick, S. 1980. *Manual for composting sewage sludge by the Beltsville aerated static pile method.* EPA -600/8-90-022. Cincinnati, OH: Municipal Environmental Research Laboratory.

7 Facility Design

INTRODUCTION

Proper design of a composting facility is imperative for cost-effectiveness, good operations, and environmental control. The design determines the capital investment and future operational costs. Some of the capital costs impacted by the design are listed below, along with various pieces of equipment. It is obvious that the extent of capital improvements, as well as the equipment needed, will vary with the feedstock to be composted and the system selected. A homogeneous feedstock such as biosolids will require less equipment than municipal solid waste (MSW). In the case of biosolids using wood chips, front-end loaders (FELs), hoppers, conveyors, mixers, agitators or windrow equipment, blowers, and screens may be all that is needed. If, on the other hand, yard waste is used at the facility, grinding equipment is also needed. MSW composting would also require ferrous metal removals, countercurrent equipment to remove nonferrous metals, plastic removal equipment, and other refining equipment. The following list indicates some of the site capital costs, as well as the equipment that may be needed:

- Site improvements
  - Roads
  - Structures such as buildings and sheds
  - Fences
  - Grading
  - Drainage
  - Utilities such as electricity and water conveyance
  - Ponds or storm water containments
  - Aesthetic requirements
  - Odor control facilities
  - Vehicle wash areas
- Equipment
  - Front-end loaders
  - Screens
  - Trucks
  - Windrow equipment or other turning equipment
  - Sprayers
  - Water conveyance vehicles
  - Mixers
  - Scales
  - Air classifiers

- Ferrous metal removers
- Countercurrent equipment
- Grinders
- Hoppers
- Conveyors
- Dust removal equipment

PROCESS FLOW DETERMINATION

In Chapter 3, various process flows were illustrated. Once the process flow is determined, a more detailed designation of the material movement, equipment needed to be determined for specific operations, and evaluation of the materials handling can be made. The material flow for biosolids and MSW will be illustrated below.

The movement of material throughout the entire process depends on the complexity of the system, type of feedstock, and its homogeneity. Simple processes having a single feedstock, such as yard waste or biosolids, may require very little materials handling steps. In the case of clean yard waste, the primary steps needed are transporting the waste to the facility, grinding, forming windrows, water application (if needed), removal of material from the windrows, screening, storing, product distribution, and removal of residuals from the site. If the feedstock is biosolids using wood chips as the bulking agent, few steps are needed. These may consist of delivery of the biosolids and wood chips to the site, mixing with a mobile or stationary mixer, building the piles over an aeration system, removal of the uncured compost to a curing area, screening either before or after curing, moving the compost from the cure area to a screen and into a storage area, and product marketing.

A combination of feedstocks or heterogeneous waste often requires numerous additional steps to those indicated earlier. Figure 7.1 shows the process flow for a sewage sludge or biosolids composting operation similar to the one designed in the ASP in Davenport, Iowa. This facility is shown in Figures 7.2 and 7.3.

In Davenport, a yard waste facility is located on the same site as the composting facility. Therefore, in addition to the steps shown in Figure 7.1, Davenport also grinds the yard waste. The first step in the materials flow is the delivery of biosolids to the facility. This can be accomplished either directly from the dewatering building through a conveyor system to a hopper located in the composting facility or via depositing into a three-sided concrete bay. In the case of the enclosed facility in Davenport, Iowa, the biosolids are trucked to the composting facility and deposited into a hopper. From a live-bottom hopper, the biosolids are deposited onto a conveyor. Bulking agents, such as wood chips or shredded yard waste, are added to the conveyor carrying the biosolids. These materials are deposited into pug mills for mixing. The ratio can be controlled either by using scaled conveyors or empirically based on the discharge rate of the biosolids or bulking agents. The mix is deposited in the composting hall, where FELs pick it up and place it over an aeration system. Following composting, the compost and bulking agent are removed and taken to the screening area. Once the compost is screened, the bulking agent is recycled, and the product to be cured is placed over an aeration system. The final cured product is either bagged or sold in bulk.

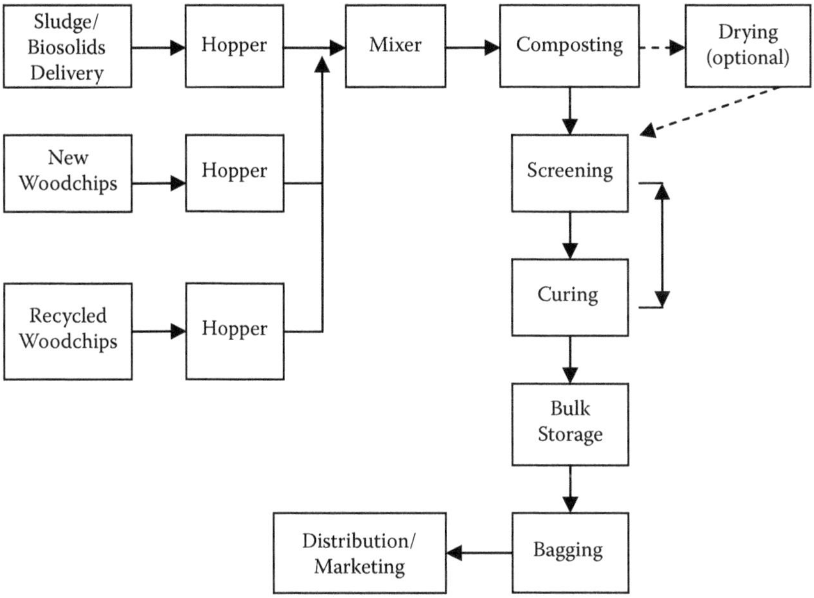

FIGURE 7.1 Process flow illustration for a large-scale sludge or biosolids ASP composting operation.

FIGURE 7.2 Davenport biosolids composting facility. Adjacent to the facility is the biofilter for odor control.

A small operation may need only a mixing box where the biosolids or other feedstocks and bulking agents are combined. This is illustrated in Figure 7.4.

In a large operation, a continuous operation may be desirable. This is accomplished using pug mills. The pug mills are fed from hoppers connected by conveyors. Either using weight belt conveyors or calibrating the delivery of bulking agent relative to the feedstock is necessary. These procedures are illustrated in Figures 7.5 and 7.6.

In Figure 7.7, an MSW process flow is depicted. Here, feedstock reception was the first activity. This phase will probably require a building with a concrete floor for

FIGURE 7.3 The composting hall in the Davenport facility.

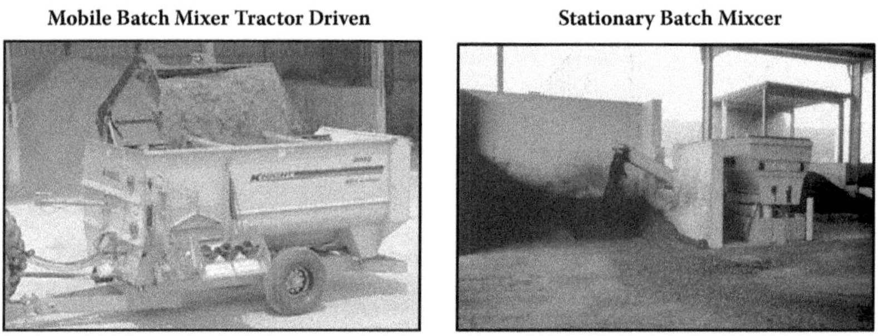

FIGURE 7.4 Batch mixers used as either mobile or stationary mixers.

vehicles to enter and dump the solid waste. On the floor FELs or bobcats will remove undesirable items, such as refrigerators, stoves, tires, automobile transmissions, and other material. Once these are removed, the material is pushed into a hopper system, located in the back end of the reception building. Alternatively, the sorted feedstock could be picked up by a grapple or FEL for deposit into a hopper.

The material in the hopper is moved by a grapple or other device, deposited onto a conveyor, and passed through a hammermill or other grinding device. After the MSW is shredded, it is deposited onto a conveyor belt, and individual items are separated.

Separation can consist of ferrous metal removal, nonferrous metal removal by countercurrent devices, and screening to remove plastics. The ferrous metals, nonferrous items, and plastics or other undesirable material are conveyed and deposited into containers. At this point, the waste is fairly homogeneous and is then conveyed into a mixer. If at this time amendments or water addition is needed, this can be provided. Following mixing, the composting and curing phases occur. Material handling depends on the system used. If an agitated bed system is to be used, an

FIGURE 7.5 Bulking agent hoppers at Davenport, Iowa.

FIGURE 7.6 Pug mills and conveyors at Davenport, Iowa.

agitator must be provided. In the United States, the primary agitated bed system is International Processing System (IPS). Longmont and OTV have also provided agitated bed equipment.

In Canada, the Edmonton facility uses a drum system, followed by an agitated bridge system provided by the Italian firm Sorain Cecchini. The Edmonton facility does not carry out initial separation other than bulky items at the tipping floor. Extensive separation occurs after the composting process and before curing. The

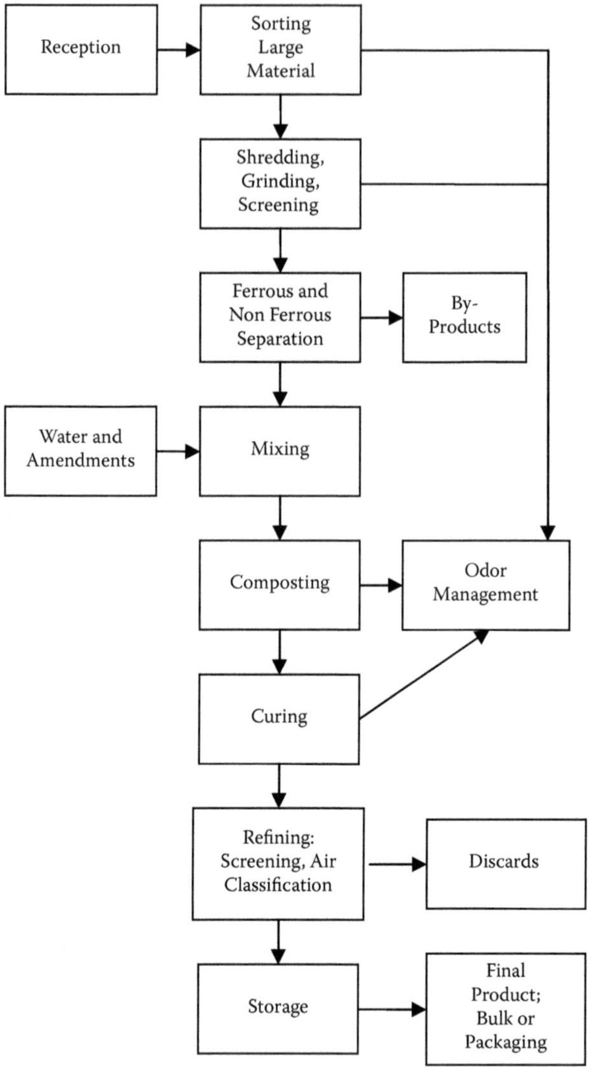

FIGURE 7.7 General MSW composting flow process.

Bedminster system, developed in the United States and used in several locations, uses the drum system. One of the largest drum systems worldwide is the DANO system. There are two different basic concepts in drum design. One concept, such as the Bedminster, uses compartments within the drum. Material is retained for twenty-four hours in the first compartment, then moves to the second, and subsequently the third. In the second concept, such as that used by DANO, Edmonton, and other drums, the waste remains in the drum for several hours before being discharged for additional composting. The drum system in Edmonton, Canada, is shown in Figure 7.8. The smaller building preceding the drums is the receiving hall, where separations of

FIGURE 7.8 Edmonton, Canada, drum composting system.

large materials take place. The MSW is then placed into the drums, where it is mixed with biosolids. Following mixing, the material is conveyed into the composting hall. Figure 7.9 shows the turning machine, which is used for composting.

The third facility illustrated is the Ngau Tam Mei horse manure composting facility located in Hong Kong. The facility is capable of handling up to 20 tonnes (22 tons) per day. This facility uses a combination of a drum system and an aerated static pile system. The process flow is similar to the MSW flow depicted in Figure 7.7. Figures 7.10 and 7.11 illustrate this facility.

The horse manure arrives in plastic bags. The manure contains bedding material primarily straw but occasionally wood chips. The bags with manure are

FIGURE 7.9 Edmonton, Canada, MSW and biosolids composting facility, showing the turning machine.

FIGURE 7.10 A Ngau Tam Mei horse manure facility in Hong Kong showing hopper and separation equipment.

FIGURE 7.11 The drum that follows the processing facility housing the separation equipment. The composting and curing bins are shown in the background.

dumped on the tipping floor and manually debagged. Any large undesirable material is removed for disposal. However, truck delivery of the horse manure in bulk is also possible. A small FEL picks up the manure and deposits it into a hopper. From the hopper, it is conveyed under a magnet to remove ferrous metals. The ferrous metals are deposited in a container for recycling. Following the magnet, the manure passes through a countercurrent device, which removes nonferrous metals. Principally, aluminum is removed and deposited into a container for recycling. A conveyor carries the manure into a disc hopper, which removes plastics. The cleaned manure is then transferred through a conveyor to a drum. Drum retention time varies with the manufacturer and the facility. Many systems worldwide keep the retention time to twenty-four hours or less. Since that short time does not accomplish composting, composting or curing needs to be accomplished by either an aerated static pile, windrow, or agitated bed technologies. In this facility, the composting and curing are done in aerated bins. The drum does a good job of mixing and initiating temperatures.

Generally, manure does not need such extensive cleaning. However, this example was chosen because it can be applied to numerous wastes, such as biowaste, contaminated food waste, and other wastes. It is therefore important to obtain the best possible information as to the incoming feedstock. A facility may also be designed with additional equipment for future application or anticipation of changing conditions.

SITE DESIGN AND IMPROVEMENTS

The planning and design of a site begin once the location is determined. The first requirement is a site plan and layout. This design must incorporate the following major aspects. There may be other items for specific locations.

1. Traffic flow and location of roads
2. Site improvements
   a. Grading
   b. Drainage
   c. Storm water containments
   d. Fences and gates
   e. Road construction
   f. Utilities
   g. Lighting
   h. Aesthetics
      i. Trees and shrubs
      ii. Grassed areas
3. Structures
   a. Location
   b. Size and type of buildings
   c. Odor control facilities
   d. Vehicle wash areas

Traffic Flow and Location of Roads

Locating the proper egress into the facility is important to avoid congestion of incoming feedstocks, supplies, and removal of unwanted waste, and for transporting the final product. A facility may therefore decide on having more than one entrance.

Site Improvements

One of the first activities is preparing the site for the composting operation. This involves grading for equipment operations, structures, traffic, and proper drainage. The construction of berms, runoff mitigation, and water containment structures for fire and storm water/sediment containments may be needed, depending on location and local regulations.

Other major site activities are location and installation of utilities, such as power, water, and sewage. The location of permanent or temporary structures includes a weight scale, office building, and possibly a feedstock receiving facility. A bagging house may be needed as well. For agitated bed and other major equipment, the structures may include bins, blower housing, and other structures as needed for the operation of the particular system. As was pointed out earlier, some systems may contain designed structures such as drums to be located outside and adjacent to the main operational facility. In these systems odor control equipment or structures, such as a biofilter, are needed. It is imperative to have a good layout of the facility initially to avoid any changes later, which could interrupt operations or result in excessive additional costs. Generally, fencing and gates need to be installed to prevent trespassing and damage to equipment.

Relatively low-cost systems such as windrow or ASP may require minimal site improvements. These may include a trailer for office and worker facilities, a staging area for feedstock delivery and bulking agents, a screen location, and any runoff containments. Ponds may be needed for fire protection or water for the composting operation. For ASP operations using negative aeration, a biofilter is needed for odor control. This is not needed where positive aeration and fabric covers are used.

As part of a design, consideration needs to be given to aesthetics. A good-appearing and clean facility will give the public, local officials, and regulators a feeling that the facility is being operated properly. The aesthetic conditions could be grassed areas, trees and shrubs, and flower beds. The use of compost in these areas is an excellent selling point. Grassed areas can be very effective in reducing excessive contamination by runoff diverted to storm water/sedimentation basins. Grassed areas are water filters.

Structures

The need for structures depends on the feedstock, system to be used, and location. Feedstocks such as biosolids, MSW, and food waste generally require more structures on-site than yard waste. Complex systems requiring proprietary equipment and contaminant separation require more structures than normal windrow or ASP operations. A composting operation located at a wastewater treatment plant may not need any structures, or at minimum, a trailer for equipment and lockers. Worker facilities are already available at the treatment plant. Similarly, a yard waste composting operation located at the landfill may also require few structures, such as a weight scale or worker facilities. Sites located relatively near residents or commercial enterprises may require structures to avoid odors and dust.

EQUIPMENT

Correct equipment selection is critical. Not only is the type of equipment important, but also its capacity. Many facilities make the mistake of selecting the wrong equipment. Thus, the result is that the equipment does not do a proper job or, in extreme cases, it is useless. This is a waste of capital costs. For example, the recommendation for large facilities that build piles is to use FELs having 6.1 to 7.6 m^3 (8 to 10 CY) bucket capacity, with either a push plate or rotating bucket. This can save considerable time in building higher piles. The primary equipment used by composting facilities is listed below:

- Front-end loaders
- Trucks
- Weight scale
- Windrow turning equipment
- Proprietary turning, agitating, or pile-building equipment
- Blowers for aeration and air distribution
- Odor control equipment
- Screens
- Bagging equipment
- Grinding equipment

A very excellent description of equipment available for composting operations can be found in Chiumenti et al. (2005). Therefore, only a brief description of most of the equipment will be given. With the exception of windrow equipment, the aeration system used in both nonproprietary and proprietary systems is the most critical. Since composting is an aerobic process, forced aeration is required for most systems. Therefore, the design of the aeration system and blowers is very important and will be described here.

The following is a brief description of some of the equipment listed above.

Front-End Loaders (FELs)

The FEL is the workhorse of a facility. It is used for moving feedstocks on tipping floors, building piles, loading trucks with the final product, and other activities. The size and capacity of the FEL should be tailored to the activity. In building piles, a large-capacity bucket should be used, e.g., 7.6 to 9.1 m^3 (10 to 12 CY). If the desire is to build high piles of 3.0 m (10 feet) or greater without traveling on the material or packing it, then buckets that can rotate, such as a high dump bucket or buckets with a push plate, are very useful.

Windrow Turning Equipment

There are numerous manufacturers of turning equipment. Basically, there are two types: straddle and side-operated machines. The straddle windrow machine traverses over the windrow. In the process, it agitates the composted material at very high speeds. There are straddle machines that also pick up the compost and place the material on a conveyor for deposit into a new windrow. Some machines have a watering spray system to reduce dust and odors. In the United States, the straddle type is most common and is shown in Figure 7.12. The side-operated units traverse along the side of the windrow and pick up the composted material. Some machines just deposit the material on the other side of the machine, whereas others deposit the material on a conveyor for deposit into a new windrow. A side-operated machine is shown in Figure 7.13. Many of these machines are well described in Chiumenti et al. (2005).

Proprietary Equipment

There are many manufacturers of proprietary equipment in the United States and Canada as well as in Europe and Asia. Haug (1993) and the Composting Council of Canada (1995) mentioned early equipment manufacturers. Other, more recent manufacturers are indicated in Chiumenti et al. (2005).

There are several systems currently in existence but not manufactured in the United States anymore, or they have not been built recently. These are:

American Bio Tech
Bedminster
Purac
Taulman/Weiss
Fairfield Service Company

FIGURE 7.12 Straddle windrow turning machine.

FIGURE 7.13 Side-operated windrow turner.

The main equipment manufacturers currently having systems or proprietary equipment in the United States are:

- Aerated static pile
  - Ag-Bag Environmental
  - CH2M Hill
  - W.L. Gore
  - Managed Organic Recycling (MOR)
  - Engineering Compost Systems (ECS)
- Windrow
  - ALLU Group
  - Brown Bear Corporation
  - Double T Equipment Manufacturing Ltd.
  - Resource Recovery Systems of Nebraska/KW Composter
  - SCARAB Manufacturing & Leasing, Inc.
  - SCAT Engineering, Inc.
  - Wildcat Manufacturing Company, Inc.
- Agitated bed
  - IPS
  - Longwood Manufacturing Corporation
- Container
  - Engineering Compost Systems, Inc. (ECS)
  - Green Mountain Technologies
  - NaturTech Composting System
  - Wright Environmental Management, Inc.
- Drums
  - A-C Equipment Services

BASIC DESIGN INFORMATION

The following section describes in detail some basic design considerations related to aeration. Other than outdoor windrow systems, forced aeration systems are the most important aspect of composting. Even agitated bed systems use forced aeration. The enclosed LRI windrow facility in Washington state uses forced aeration. In Denmark, there is a facility with an outdoor windrow system having an aerated trench with negative aeration for odor control, as well as providing additional aeration.

Aeration Systems

An aeration system consists of the aeration pipe system under piles, the manifold, and blowers. A major consideration in the aeration system is the head loss, as that will determine the blower size and, consequently, the horsepower and energy uses.

The basic aeration pipe systems are:

- Pipe
    - Plastic
        - Rigid
        - Flexible
    - Iron
- Trench or channels
- Spigot
- Aerated block

Pipe is the most common system being used by small facilities and some large ones (Figures 7.14 to 7.16). These are used in open, covered, or enclosed ASP systems. Most of the pipes are plastic, either with predesigned uniform holes or with slots, or designed holes. In the latter the holes are spaced to provide more uniform aera-

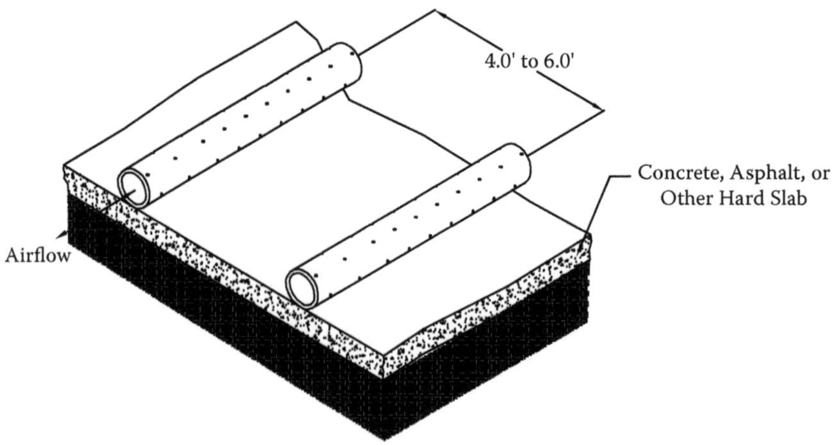

FIGURE 7.14 Aboveground standard plastic pipe designed with perforations.

FIGURE 7.15 Aboveground perforated pipe assembly system, showing manifold and blowers.

FIGURE 7.16 Aboveground pipes to provide aeration for ASP, showing pipe placement in extended pile.

tion depending on pipe length. Pipes need to be spaced properly to provide uniform aeration. Generally, pipes are spaced between 90 and 120 cm (3 and 4 feet) apart.

Perforated iron pipes have also been used. These can be recovered and reused. Injury to workers has occurred with iron pipes. Blower capacity depends on number of pipes, pipe length, and head loss through the system. The head loss is a function of pile height, pipe configuration, and whether negative or positive aeration is used.

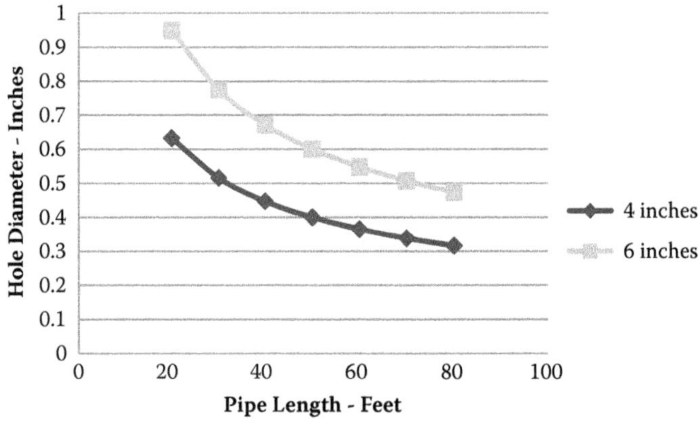

FIGURE 7.17 Hole diameter spaced at 6 inches as a function of pipe length for 4- and 6-inch-diameter pipes.

Typically 10 or 15 cm (4 or 6 inch) diameter pipes are used. Many times drainage-perforated pipes with holes predrilled are used. There are two options available. One is to vary the size of the hole and maintain a uniform hole spacing. The other is to have a uniform hole diameter and vary the spacing. Figure 7.17 shows the hole diameter spacing in the pipe at a distance of 15 cm (6 inches) apart. This functions for pipes of 10 and 15 cm (4 and 6 inch) diameter. The general formula is

$$\text{Hole diameter} = \sqrt{D^2 \times S/L \times 12}$$

where
D = pipe diameter (in.);
L = pipe length (ft.);
S = hole spacing (in.)

In covered facilities, trenches or channels and spigots have been used. Perforated blocks have also been used for both aeration and biofilters to distribute the air. Figures 7.18 and 7.19 show trench systems. Either holes or slots have been used. Figures 7.20 and 7.21 show the spigot system, and Figures 7.22 and 7.23 show the block system. In the spigot system, a pipe is laid beneath the floor with spigots rising to the surface. The spigots are placed at intervals to provide proper aeration. Block systems are usually proprietary.

Aeration Requirements

The aeration requirement is necessary for three functions:

- Microbial metabolism
- Temperature control
- Moisture control

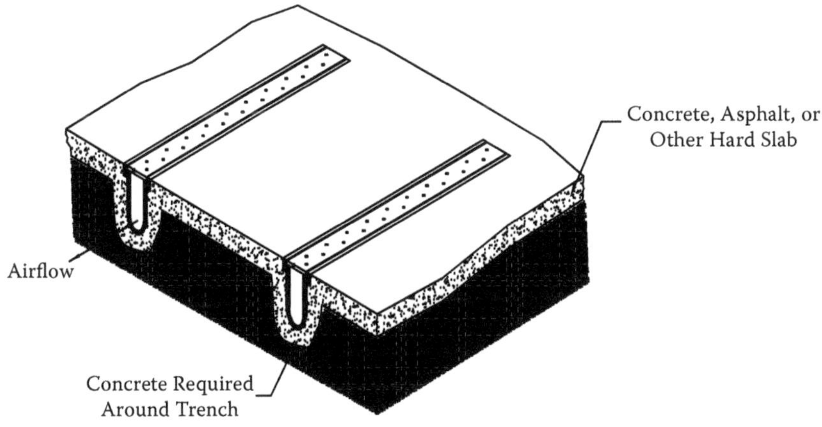

FIGURE 7.18 Design of trench system.

FIGURE 7.19 Trenches or channels used for supplying air to piles.

Microbial metabolism, resulting in decomposition of the organic matter into a stabilized product, requires oxygen. In forced-air systems such as ASP, and to some extent in most enclosed systems, blowers provide oxygen. For windrow systems, oxygen is provided through convective air and turning. Haug (1993) provides stoichiometric oxygen calculations that depend on the composition of the material, and the extent of decomposition during composting, as expressed in terms of biological volatile solids (BVS). However, from a practical operational viewpoint, aeration rates for microbial decomposition are expressed as cubic meters per hour per dry tonne (cubic

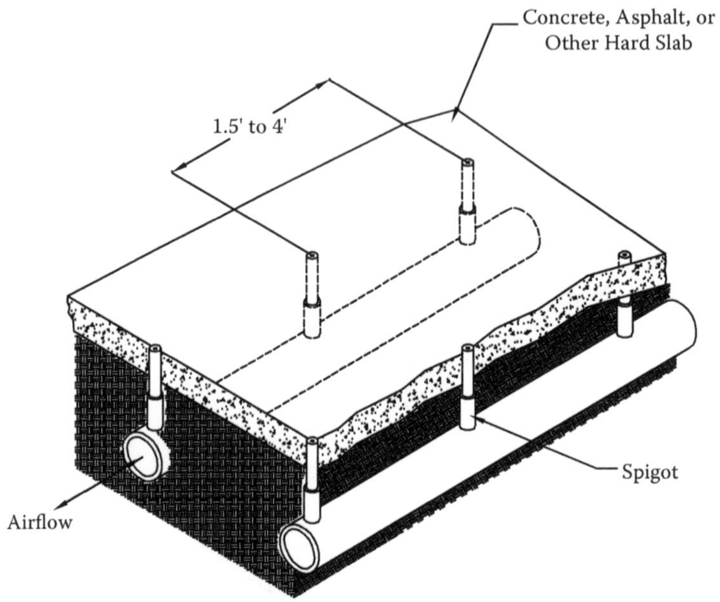

FIGURE 7.20 Diagram showing spigot design for aeration.

FIGURE 7.21 Spigots placed for aeration.

feet per hour per dry ton—cfh/dt). Generally, if the bulking agent is relatively inert, e.g., sawdust, wood chips, or other slowly or nondegradable material, the aeration rate is based on the feedstock. For example, in the case of biosolids, the aeration rate is based on m^3/dry tonne (cfh/dt) of the biosolids only. Empirically, it has been found that to provide oxygen for biological activity, the aeration rate ranges from 6.23 to 15.6 m^3/h/dry tonne (200 to 500 cfh/dry ton).

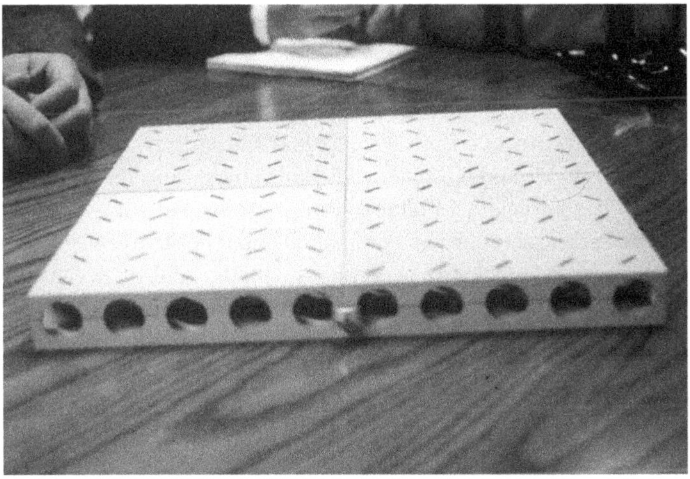

FIGURE 7.22 A perforated block used for aeration floors.

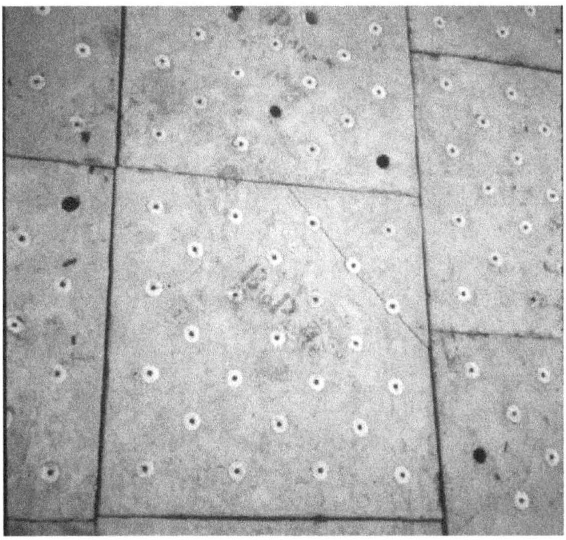

FIGURES 7.23 Blocks used to supply air to ASP and biofilters.

However, during decomposition, energy is evolved, and temperature rises from ambient through mesophilic to thermophilic. Therefore, the aeration system needs to control the temperature and manage the process, including heat removal. Furthermore, moisture removal is often necessary in order to be able to process the product. Equipment needed for particle size determination and removal of contaminants best operates at moisture levels below 45%. However, it is best to keep the compost moisture between 40 and 45% to avoid dust. Excessive wet feedstocks will reduce the effective pore space and the availability of oxygen to the matrix.

The rate for temperature control has the greatest demand, and therefore this rate is used to design the aeration system. Empirical data have shown that between 90 and 160 m^3/h/tonne (3,000 and 5,000 cfh/dt) is appropriate for both temperature control (heat removal) and moisture control. Typically, moisture control is practiced toward the end of the process. Both Haug (1982) and Murray and Thompson (1986) indicated that, for sludge and biosolids with wood chips, 11 to 160 m^3/h/tonne (3,800 to 5,000 cfh/dt) was sufficient for temperature control.

As indicated earlier, aeration can be supplied under positive or negative air. When aeration is supplied under negative aeration, as with the ASP, there can be condensation. This condensate must be removed; otherwise, it can accumulate in the ductwork. Drainage must be provided.

Aeration can be either intermittent or continuous at low rates. At intermittent rates, the on-off sequence must be short to avoid depletion of oxygen. In Figure 7.24 oxygen is depleted within twenty minutes when a blower is shut off. Anaerobic condition in a pile can occur when pile oxygen is less than 5%. Typically, oxygen levels should range from 10 to 18% in all systems. In Figure 7.25, oxygen became depleted in sixty minutes in the windrow following turning. Convective air maintains the proper aeration if there is sufficient porosity.

CONCLUSION

The design of a composting facility is primarily a function of the type of system and its complexity. Location can also affect design. Proximity to receptors often requires enclosures and odor control. The primary design components are site preparation, feedstock handling, system selection and installation, and odor control.

Outdoor windrow or aerated static pile (ASP) systems are the least complex, and the design is relatively simple. For windrows, the primary design features contain

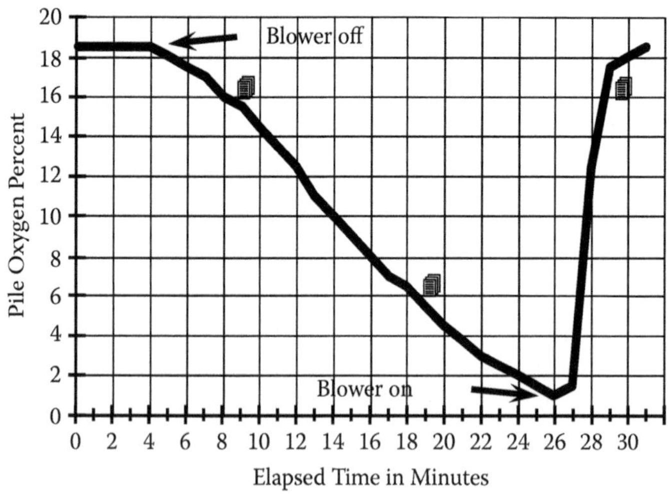

FIGURE 7.24 Oxygen depletion rate in an ASP operation when a blower is shut off.

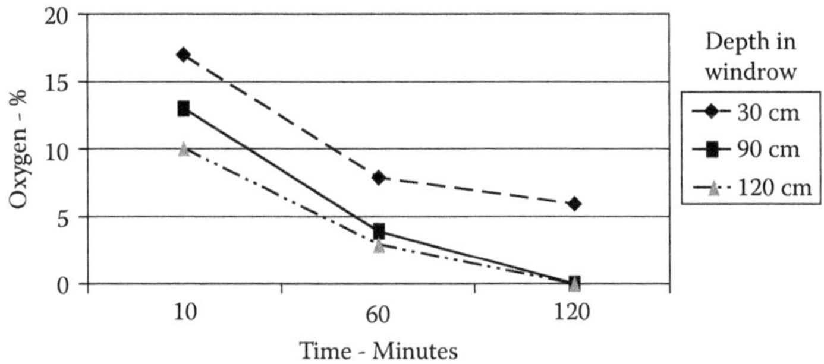

FIGURE 7.25 Oxygen depletion rate during windrow composting.

the feedstock handling facility, composting site, and site water management. In that case, the truck bringing in the feedstock discharges it on a flattened portion of the windrow. The windrow machine subsequently mixes the material. Site preparation is the greatest part of the design. Equipment generally consists of front-end loaders (FELs), windrow equipment, and a screen. No special aeration equipment or odor control equipment is necessary.

ASP systems are slightly more complex in design. Blowers, pipes, and temperature control devices are usually needed. Therefore, power to the site is usually brought in. However, sites have used generators. A feedstock handling system, consisting of a mixer, is often provided. FELs are used to build the piles. Once piles are built, they typically are not moved or mixed for twenty-one to twenty-eight days. After that time, the compost is either screened or cured. If sufficient space is available, curing can be in place, and thus there is no requirement for moving the piles. A screen is needed for particle sizing and removal of contaminants. Under negative aeration, biofilters are usually used for odor control.

Agitated bed systems require a building to house the equipment. In addition, the design requires supporting structures for the agitated bed equipment. These systems typically use conveyors and FELs for material handling. Either mixing is done with mixers, or the feedstock is mixed with the bulking agent directly in the bed. Odor control is needed and usually provided by a biofilter. A screen is usually needed for particle sizing and removal of contaminants.

Container systems usually require the least site preparation. As with agitated bed and ASP, mixing of the feedstock with a bulking agent is usually accomplished with a mixer. Depending on the length of time the material remains in the containers, space and facility for curing may be needed. As with the previous systems, odor control and screening are needed.

Drum systems do not accomplish composting. The drum does an excellent job of mixing. Some drum systems do accomplish minimal composting by achieving an increase in temperature, as a result of the retention time in the drum. The longer the retention time, the more composting that is achieved, but also the capacity of the system is reduced. Conveyors and FELs are usually needed for further material handling after the material is discharged from the drum.

With the exception of the windrow system, some aeration is required for oxygen provision, temperature control, and moisture control. Aeration system design involves conveyance of the air through pipes, channels, or spigots; blowers with piping and valves; and a control system through a feedback or manual operation.

REFERENCES

Chiumenti, A., Chiumenti, R., Diaz, L., Savage, G. M., Eggerth, L. L., and Goldstein, N. 2005. *Modern composting technologies*. Emmaus, PA: BioCycle.

The Composting Council of Canada. 1995. *Composting technologies and practices*. Ottawa, Ontario, Canada: The Composting Council of Canada.

Haug, R. T. 1993. *The practical handbook of compost engineering*. Boca Raton, FL: Lewis Publishers.

Murray, C. M. and Thompson, J. L. 1986. Strategies for aeration pile systems. *BioCycle* *27*(6) 22.

8 Economics

INTRODUCTION

The economics of composting is highly variable. Site conditions or specific technologies require building and enclosures that result in much higher capital costs than those sites that are remote and the technology can be operated in the open. The cost to public agencies, such as municipalities or districts, would be different for private companies. A major aspect for public agencies is the avoided costs.

The economics of composting depend on a number of variables. The traditional categories are:

- Operational and maintenance (O&M) costs
- Capital costs
- Revenues

Capital costs will depend on the system or technology selected. In general, the order of increased costs is listed below:

- Passive aerated pile
- Windrow
- Aerated static pile
- Within vessel

Even within these categories, there will be variability in costs. The size of the operation will determine the types of equipment and number.

Location will determine such costs as land, structures, odor control, roads, and other site costs. Ancillary costs relate to leachate disposal, runoff collection facilities, waste disposal, and residuals disposal. These may include not only site preparation costs, but also hauling costs, which will depend on distance.

Another consideration, especially for large facilities, is labor vs. equipment. Labor costs and fuel are the two variables that will escalate with time. Product marketing is an important financial consideration. Not only does it provide income to a facility, but it also incurs costs.

ECONOMICS

As indicated earlier, the economics of composting will vary, whether it is a public agency or a private company establishing and operating a facility. A public agency has several choices:

- Owned and operated by the utility or public entity
- Owned by public entity and privately operated
- Contracted for privately owned and operated

Within these categories, there are other considerations. For example, should the product marketing be contracted out to a private company, even though the facility is owned and operated by the public utility?

Avoided costs are primarily a consideration for public entities.

Avoided Costs

Avoided costs are those costs a public agency may incur currently or potentially may need to manage organic wastes and to purchase organic products for parks, public work projects, etc. The avoided costs may include disposal or recycling fees, transportation costs, equipment maintenance costs, management costs, and associated utility expenses. Another aspect of the avoided costs is the value of products, such as compost or mulch, that would not have to be purchased. The following illustrates some of the avoided costs. These are not necessarily all the associated costs.

- Disposal costs
  - MSW
    - Transportation
    - Labor
    - Landfill fees or costs
      - O&M costs
        - Labor
        - Utilities
        - Monitoring
        - Road and site maintenance
        - Hauling
        - Vehicle and equipment maintenance and depreciation, including washing
  - Sludge or biosolids
    - Other treatment costs—for a comprehensive full-cost accounting, see WERF (2003)
      - Land application
        - Transportation
        - Application—equipment and labor
        - Permitting and monitoring
        - Truck washing
        - Equipment maintenance costs
      - Sludge anaerobic or aerobic digestion costs
        - Facility costs
        - Operating costs
        - Management of residuals
        - Energy capture and saving
      - Incineration

        - Equipment replacement and maintenance costs
        - Monitoring
        - Labor
        - Fuel
        - Ash disposal
        - Building maintenance costs
        - Utilities
      - Landfilling (see MSW costs above)
  - Yard waste
    - Chipping and grinding
      - Labor
      - Equipment maintenance and depreciation
    - Hauling costs to another facility
- Product use in public works, parks, cemeteries, general landscaping, etc.
  - Mulch
  - Soil amendments
  - Savings in water usage

Steps Needed to Arrive at the Cost of a Facility

Before arriving at the estimated cost of a facility, a conceptual design needs to be done. The following is an example for an open ASP system designed to handle a large quantity of biosolids:

- Identification of the technology, such as ASP
- Perform a basis of design
  - Daily materials balance
  - Proposed process flow
  - Based on materials balance size, each of the areas in the process flow
  - Determine type of equipment for material movement, e.g., FELs, conveyors, trucks
- Preliminary site plan
  - Layout location of roads, office, and weight station
  - Identify electrical system for composting, curing, and screening
  - Layout location of structures and size of structures based on material balance
  - Layout location of ASP pad and size based on material balance
  - Layout location of curing area and storage areas based on material balance
  - Locate screen
  - Locate and size biofilter
  - Locate water control aspects
- Define feedstock and bulking agents or additives receiving
  - Buildings or structures
  - Equipment to move material
  - Storage, e.g., hoppers

- Mixing system
  - Type of system
  - For a large system, number and capacity of hoppers and conveyors
  - Capacity of feedstock based on delivery and operating time
  - Capacity of bulking agents
  - Volume of discharge of mix as related to operating time
- Composting
  - Preliminary design of composting pad, including location of blowers
  - Size blowers based on pile volumes
  - Locate control system for blowers
  - Aeration system
    - Disposable
    - Reusable
    - Trench
    - Block
    - Other
  - Design biofilter, including irrigation system
- Curing
  - Determine if curing is going to be aerated or static
  - If aerated, locate and size blowers; type of piping
- Screening
  - Locate and size screens; determine type of screen
  - Identify movement of materials, both recycled bulking agent and finished product
- Storage
  - Identify location of both bulking agents, recycling bulking material, and finished product
  - If it is determined to have a bagging operation, size and identify equipment
- O&M cost estimate
  - Personnel requirements—hourly costs, hours per day, full-time equivalents
  - Equipment operation—hauling costs for feedstock and bulking agent
  - Fuel
  - Electrical costs
  - Equipment maintenance and replacement, both large and small
  - Water and sewage disposal
  - Laboratory fees and supplies
  - Permitting costs and legal fees
  - Security
  - Administration
  - Janitorial
  - Bulking agent replacement
  - Biofilter maintenance
  - Site maintenance, including lighting
  - Monitoring, permitting

- Public relations and legal costs
- Contingency
- Feedstock fees and revenue
- Compost sales and revenue

Preliminary Costs

If a site is owned or to be purchased, the first costs are permitting and the associated legal costs. This may require an environmental assessment or even an environmental impact report that could require public hearings. This may be very expensive. The permitting agency or agencies may require a preliminary design and the environmental impacts of potential odors, noise, and traffic. In most cases, the implementation of an environmental assessment will require hiring a consulting engineering firm. During the permitting process, it is advisable to begin public relations. This may involve meetings with the community, producing and distributing a newsletter, and presentations at local groups such as the Lions Club or other organizations.

Compost Facility Capital Costs

The compost facility capital costs will vary considerably with the feedstock (biosolids, yard waste, food waste, MSW), volume to be handled, location, and site requirements. Once a site is located and permits obtained, the capital costs associated with building the facility are:

- Site development costs
  - Roads
  - Preliminary design
  - Site work
    - Grading
    - Excavations for building foundations and footings
    - Runoff, storm water, leachate collection, and ponds
    - Fencing and other security needs
    - Pavements
    - Weigh station
    - Office
- Process development costs
  - Mixing area
    - Building
    - System
    - Feedstock receiving
    - Bulking agent storage and handling equipment (if needed)
  - Composting area (depends on technology selected); the following is for an ASP system:
    - Pavement
    - Aeration

    - Blowers
    - Piping
    - Aeration system or floor
  - Odor control
    - Biofilter
      - Blower
      - Piping
      - Irrigation system
    - Meteorological station
  - Electrical
  - Control systems
    - Mixing
    - Aeration
    - Odor
  - Curing area
    - Pavement
    - Blowers
    - Piping
  - Product storage
    - Pavement
- Screening and refining area
  - Pavement
  - Electrical
  - Control
  - Equipment
- Moving equipment costs
  - Front-end loaders
  - Vehicles

Table 8.1 shows an estimated capital cost for an open-air ASP system in the year 2000 designed to handle 250,000 tons of biosolids per year on a permitted site. This information presents only the relative cost and primary capital cost required.

Operational and Maintenance Costs (O&M)

Labor

Labor requirements will differ, depending on the size of operation, technology, and automation. Not all personnel categories or equipment types would be needed in every case. The principal labor categories required are:

- Managerial
- Supervisor
- Equipment operators
- Mechanics
- Clerical

TABLE 8.1
Primary Estimated Capital Costs in the Year 2000 for an ASP Open-Air System Handling 226,850 Tonnes (250,000 Tons) of Biosolids

Item	Unit Cost	Quantity	Estimated Cost ($)	Subtotals and Totals
Site work			$3,014,928	
Compost aeration			$1,205,433	
Mixing building and system			$1,893,059	
Odor control			$2,361,731	
Electrical			$635,636	
Composting control system			$130,000	
Mixing control system			$25,000	
Biofilter control system			$45,000	
Construction subtotal				$9,110,790
	General Conditions			
Mobilization	2%		$18,236	
Contingency	15%		$1,369,354	
Contractor overhead and profit	20%		$2,096,276	
Performance bond				
Engineering	10%		$1,136,224	
Subtotal				$4,620,090
	Equipment			
Front-end loaders	308,000.00	4	$1,232,000	
Screen	$250,000.00	2	$500,000.00	
Power washer	$6,000	2	$12,000	
Subtotal				$1,744,000
Project total				$15,474,880

- Laboratory assistant
- Compost marketing individual

Other O&M Expenses

- Building and other structure maintenance
- Large equipment maintenance and spare parts—FELs, windrow machines, screens, mixers, etc.
- Small equipment maintenance—blowers, piping, valves, etc.
- Consulting
- Permitting and fees
- Training
- Bulking agents replacement
- Fuel
- Electricity or generator maintenance
- Water

- Supplies and tools
- Legal services
- Laboratory services, including sampling and monitoring
- Taxes
- Site maintenance—pad, fencing, road repairs, lighting, storm water maintenance, truck wash, etc.
- Control equipment repairs, replacement, and maintenance
- Waste disposal, including septage or sewage
- Biofilter—media replacement, irrigation equipment maintenance, blower maintenance

Table 8.2 shows the estimated personnel costs for a facility assuming a 22.70-tonne (25-dry-ton) operation with 23% solids based on a five-day operation of an ASP open system. Other O&M costs for the facility would be the electrical costs. These would be based on items such as hoppers, conveyors, screens, compost fans, biofilter fans, curing fans, and lighting. The cost categories include horsepower, cost per kilowatt-hours (KWH), KWH per day and year, and annual costs. In addition, there will be other O&M costs, such as:

- Hauling bulking agents
- Biofilter media replacement
- Blower and fan maintenance
- Laboratory analysis costs
- Administrative costs
- Vehicle maintenance
- Janitorial

TABLE 8.2
Personnel O&M Estimated Costs Analysis Assuming a 22.7-Dry-Tonne (25-Dry-Ton) Operation with 23% Solids Based on a Five-Day Operation

Personnel	Cost per Hour	Hours per Day	Full-Time Equivalents	Annual Cost	Subtotal Costs
Managerial/supervisor	37.50	8.0	1.0	78,000	
Equipment operators	32.00	8.0	4.0	266,240	
Mechanic heavy machinery	32.00	8.0	1.0	66,560	
Plant maintenance mechanic	32.00	8.0	1.0	66,560	
Clerical	20.00	8.0	1.0	41,600	
Laboratory assistant	25.00	8.0	1.0	52,200	
Compost marketing	30.00	8.0	1.0	62,400	
Total costs					711,048
Cost per tonne					58.80
Cost per dry ton					64.48
Cost per wet tonne					13.43
Cost per wet ton					14.79

- Miscellaneous
- Bulking agent

Potential revenue is realized from compost sales and any tipping fees.

Another example of O&M costs for an MSW facility in 2000 dollars is illustrated in Table 8.3. All these examples are provided as indications of the cost categories. Each facility must develop its specific costs. These will vary whether it is an agency, municipality, or private individual operating a facility.

Cost Comparison between Technologies and Systems

It is difficult to provide exact costs for different systems. However, the following information that was obtained from a feasibility report to the City of Palo in 2008 provides some idea as to costs of various systems (City of Palo Alto, 2008). Vender-provided costs are usually for general information and would be different in the bidding process, depending on the specific design. Furthermore, these costs do not provide land costs and site development costs.

The windrow system was the least costly. The conventional equipment cost for an operation on city-owned land was $800,000 for a front-end loader, scarab windrow turner, grinder, and trammel screen. No structures were indicated. This author recommends that a mixer and a partially or totally enclosed feedstock receiving area be constructed, especially if biosolids or food waste are to be composted.

The Ag-Bag and covered aerated pile were also considered. The costs received were budgetary costs, which differ from bid costs. The equipment included cover system, automated controls, aeration equipment, and biofilter hardware. The approximate costs were $700,000. In addition, a front-end loader, grinder, and trammel screen would be needed for an estimated cost of $625,000. The total estimated cost would be $1,325,000.

The third system evaluated was containerized, such as Engineered Compost Systems (ECS). The budgetary quote received in June 2007 was $8.9 million, including building costs to handle 30,000 tons of waste.

The site where the facility is located could greatly dictate the selection of the system and cost. Some of the proposed sites would be near residences, and only an ASP covered or containerized system would make sense, as odors would have to be contained.

Earlier it was indicated that marketing of the compost could greatly reduce costs, especially operating costs. Table 8.4 shows data that the City of Palo Alto, California, obtained from several composting marketing entities (City of Palo Alto, 2008).

ECONOMICS OF PRODUCT MARKETING AND SALES

The marketing of compost is an important revenue source, as shown in Table 8.4. For public utilities, it can significantly reduce the operating costs. Davenport, Iowa, not only receives revenue from the sale of compost, but also receives revenue from tipping fees from the disposal of yard waste. Private industry obtains revenue from disposal fees and sales of compost. Product marketing incurs expenses, as well as potential

TABLE 8.3
Illustration of the Operating Cost for an MSW Composting Facility in Year 2000 Dollars

Operating and Maintenance Costs

Labor	Rate/Hour ($)	Number of Personnel	Fringe Benefits	Annual Cost ($)
Supervisor	27.50	1	35%	75,816
Operators	17.00	4	35%	179,712
Laborers	12.00	2	35%	67,392
Subtotal cost				322,920

Material Management Costs

	Cost/tonne (cost/ton)	Tonnes (tons)	Cost ($)
Feedstock transportation	11.02	23,603	260,000
	(10.00)	2(26,000)	(260,000)
Residual disposal—5% of input	22	1,180	26,000
	(20)	(1,300)	(26,000)
Subtotal costs			286,000

Business Management Costs	Cost ($)
Legal, regulatory, consulting	10,000
Administration and billing	5,000
Association, public relations, outreach	3,000
Training, safety	3,000
Marketing, feedstock, and product @ $55/tonne ($50/ton)	26,000
Subtotal costs	47,000

Processing Costs	Cost ($)
Fuel—FELs and screen	78,000
Electric—mix box	17,500
Electric—biofilters	210,000
Electric—blowers	25,000
Electric—misc.	20,000
Water and sewer	24,000
Natural gas	5,000
Maintenance—3% capital—blowers, control system, scale $350,000	10,500
Maintenance—5% capital—FELs, mix boxes, screen $890,000	53,750
1% capital—structures, pad, electric, biofilter $2,751,500	27,500
Grounds	14,000
Operating consumables	7,000
Laboratory and misc.	25,000

TABLE 8.3 (Continued)
Illustration of the Operating Cost for an MSW Composting Facility in Year 2000 Dollars

Processing Costs	Cost ($)
Miscellaneous	5,000
Subtotal costs	522,250
Subtotal	1,178,180
Management and operating fees—20%	235,634
Contingency—5%	70,691
Total costs	**1,484,505**

Source: Courtesy of the City of Palo Alto, 2008.

TABLE 8.4
Prices of Compost Sold from Several Locations in California

Company	County	Price m^3 (per yard)	Price tonne (per ton)
Jepson Prairie Organics (Vacaville)	Solano	5 ($7)	13.35 ($14.70)
Z-Best Compost Products (Gilroy)	Santa Clara	7 ($9)	17.16 ($18.90)
South Valley Organics (Gilroy)	Santa Clara	7.6 ($10)	19 ($21)
BFI Newby Island Compost Facility (Milpitas)	Santa Clara	7.6 ($10)	19 ($21)
Grover Landscape Services (Modesto)	Stanislaus	9 ($12)	23 ($25.20)
City of Palo Alto	Santa Clara	10 ($13)	24.78 ($27.30)

Source: Courtesy of the City of Palo Alto, 2008.

revenue. It is very crucial that the marketing personnel know the product and its limitation. There have been several cases where a marketer or the utility did not understand or know the product characteristics and its use. As a result, damages incurred to plants, resulting in lawsuits or unusual expenses for replacement of the crop.

The expenses associated with compost marketing are indicated below. Not all of these categories apply in all cases. The larger the facility and the more product it produces, the more applicable are these categories.

- Labor expenses
  - Market research
  - Product research and development
  - Competitor analysis
  - Development of product literature
  - Development of product strategy

    - Product registration
    - Product demonstration
    - Product marketing
    - Sales
    - Indirect labor
    - Other
- Other expenses
    - Package design
    - Product literature
    - Communication
    - Advertising
    - Promotions
    - Trade shows, exhibits, conferences
    - Commissions
    - Transportation
    - Other expenses
- Special product production
    - Building and associated cost of HVAC, electricity, water, etc.
    - Housekeeping
    - Specialized equipment
        - Hoppers
        - Blenders
        - Conveyors
        - Dust control
        - Bagging machines
    - Labor
        - Product blending
        - Product packaging and bagging
        - Product storage
        - Indirect labor
        - Equipment maintenance and spare parts
    - Bagging and packaging supplies
    - Additives
    - Laboratory services

Product diversification can significantly improve the value of the compost and provide income toward facility operations. PayGro, a company in Ohio operated by Carl Kipp Jr., former president of the U.S. Composting Council, made diversified compost products during the 1970s. Carl produced numerous specialty products for specific markets, e.g., one for the violet flower. These specialty products were sold at higher prices than the regular compost. The City of Palo Alto, California, also produced specialty products, as shown in Table 8.5.

A more complete discussion on the utilization and marketing of compost is in Chapter 16.

TABLE 8.5
Several Products Containing Compost and Their Value as Produced by the City of Palo Alto, California

Product	Composition	Use	Price m³ (per yard)	Price tonne (per ton)
Palo Alto compost	Compost	Soil amendment, top dressing or feedstock for soil blending	7.6 ($10)	19 ($21)
Soil conditioner	Compost, wood fines, and sandy loam	Soil additive	15 ($20)	38 ($42)
Top soil blend	Compost and sandy loam	Soil amendment	20 ($26)	50 ($54.60)
Potting mix	Wood fines, compost, sandy loam, lava fines	Potting soil	20 ($26)	50 ($54.60)

Source: City of Palo Alto, *The City of Palo Alto Compost Facility Feasibility Report*, Public Works Department, Palo Alto, CA, 2008.

CONCLUSION

The economics of composting is highly variable. It depends on location, type of system, the need for structures, operating costs, and marketing.

Site design will depend on the system selected. A windrow system will require a larger site than ASP. On the other hand, ASP will require power for blowers. Odor management and control are more effective and easier with ASP or within-vessel systems.

O&M costs are also highly variable. Labor costs for ASP and some within-vessel systems could be less than those for windrow systems.

The market is also a function of location. Markets close to urban or suburban areas usually provide a higher value for compost. Bagged compost can be transported large distances and bring in higher revenue for compost. It also incurs a cost associated with the bagging operation. Diversified compost products can provide significant revenue for a facility.

Prior to establishing a composting operation, consideration must be given to the location and system in order to determine potential costs.

REFERENCES

City of Palo Alto. 2008. *The City of Palo Alto Compost Facility feasibility report.* Palo Alto, CA: Public Works Department.

WERF. 2003. *Full cost accounting protocol for biosolids management*, ed. E. Epstein. Alexandria, VA: Water Environment Research Foundation.

9 Odor Management Basic Concepts

INTRODUCTION

Odor management is essential for successful composting of biosolids, yard waste, food waste, manures, municipal solid waste (MSW), and other organic feedstocks. Odors have been the most important criterion in opposition to the establishment of composting facilities and have resulted in closure or legal action against facilities.

Composting is the biological decomposition of organic matter under controlled conditions. By-products of composting include heat, carbon dioxide, water vapor, and certain odorous compounds. The malodorous compounds emitted during composting are volatile emissions generated from chemical and microbiological decomposition of the organic matter. Aerobic decomposition is the biochemical decomposition of organic material in the presence of oxygen. Under aerobic composting conditions, the odorous compounds produced are often less pungent and offensive than those produced under anaerobic (in the absence of oxygen) conditions. The ideal condition for aerobic decomposition to occur is when hydrocarbons and oxygen are the only compounds present. Other compounds, such as sulfur and nitrogen, are present in organic matter and react naturally to form sulfur dioxides and oxides of nitrogen. Sulfur dioxides tend to have a pungent odor. During compost operations, oxygen is maintained by forcing air through the piles, natural convection, or turning the organic material to keep the biodegradation process in the less offensive aerobic form. Composting operations that tend to keep the compost material aerobic generally result in less offensive odor conditions.

Decomposition that takes place in the absence of oxygen (i.e., anaerobic decomposition) produces odors that are more pungent. Air is naturally or mechanically compressed out of the material and is replaced by methane gas, carbon dioxide, and sulfur compounds. The most notable of these compounds is hydrogen sulfide with its characteristic rotten egg odor. Odors due to anaerobic decay are generally the odors of concern when handling organic waste material.

The major sources of odors at composting facilities are the delivery and handling of raw feedstocks, active composting, curing, and storage of the finished product. The type of feedstock handled determines the type of odors generated. For example, raw sewage sludge is more odorous than digested sludge or biosolids. Grass is typically a significant source of odor at yard waste facilities. Fish wastes and certain vegetable wastes are more odorous than food processing wastes. Technology selection and facility design also affect the types of odors that are generated, quantity released, and how odors are released.

Producing a stable product is key to producing a malodor-free material suitable for marketing and use.

The selection of the appropriate technology is important, particularly in relation to potential impacts to receptors. Climate, microclimate, and geography can affect the design of a facility. Once the technology is selected, the design can have a significant impact on the release and dispersion of odors.

Optimization of the composting process requires an understanding of the fundamentals of composting (Haug, 1993; Epstein, 1997). An essential part of odor management, especially as related to treatment and control, is the identification and characterization of odors and predicting their transport and dispersion.

These aspects will be discussed in much detail in subsequent sections of this chapter.

Today, there is no reason that, through either operations, technology selection, or facility design, malodors cannot be controlled. The term *malodors* is used because we live in a world where smell and odors prevail.

This chapter will consist of the following sections:

- Assessing and evaluating odors
- Odors and receptors
- Production of odors as related to technology
- Odor management

In the section on assessing odors, I will not go into extensive detail as to sampling and analytical methodologies. I will, however, provide sources for those individuals interested in details.

The following are some additional valuable references:

Shusternan, D., Critical Review: The Health Significance of Environmental Odor Pollution, *Arch. Environ. Health*, 47, 76–87, 1992

Cain, W. S., and Cometto-Muñiz, J. E., *Health Effects of Biosolids Odors: A Literature Review*, Water Environment Research Foundation, Alexandria, VA, 2004

Harrison, E. Z., *Compost Facilities: Off-Site Air Emissions and Health*, Cornell Waste Management Institute, Ithaca, NY, 2007

Mahin, T. D., ed., *Control of Odors and Emissions from Wastewater Treatment Plants*, Water Environment Federation, Alexandria, VA, 2004

ASSESSING AND EVALUATING ODORS

BASIC CONCEPTS

What makes an odor during composting? Odorants result from the decomposition of organic matter predominantly containing sulfur and nitrogen compounds. Anaerobic conditions generate more intense, unpleasant odors. Aerobic conditions can also create odors.

How do we smell odors? See Figure 9.1.

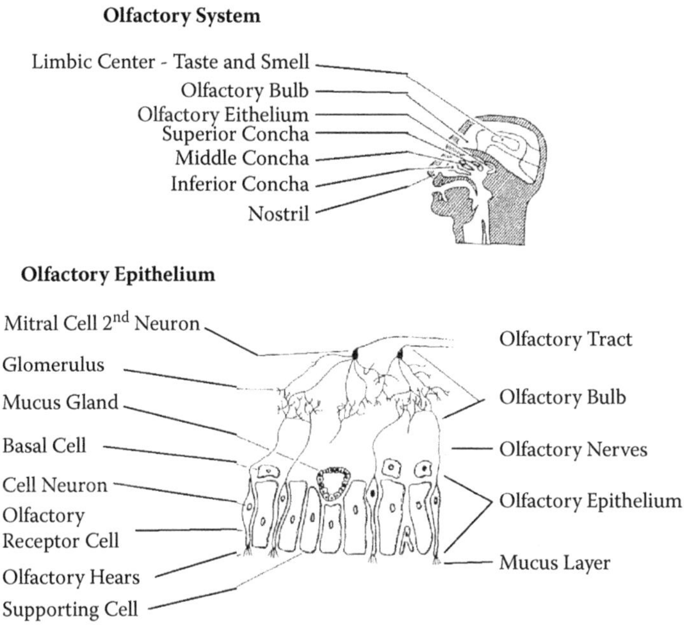

FIGURE 9.1 Odorant receptors and the organization of the olfactory system. (From Cain, W. S. in *U.S. Composting Council's 14th Annual Conference and Trade Show*, Albuquerque, NM, 2006. With permission.)

- We inhale air.
- Ten percent pass under the olfactory organ, the epithelium.
- Twenty percent pass under the epithelium during sniffing.
- There are 10 to 25 million olfactory cells in the epithelium.
- The mucus layer on the epithelium traps chemical odorants that are water soluble.
- An electrical response is created that, depending on its amplitude (strength), is sent along to the brain in the form of a pain stimulus.

How is odor perceived?

- Odor is experienced differently by different people.
- Odor can trigger memories and associations.
- Odors can trigger a pain response, e.g., irritation.

The sequence of sensory effects due to an increase in odorant concentration is shown in Table 9.1. The relationship of odor concentration on eye and nose irritation is shown in Figure 9.2.

Notice that odor concentration needs to increase beyond some point before irritation takes place.

TABLE 9.1
Effect of Odor Concentration on Sensory Effects

Concentration	Level	Effect
↓	1	Odor detection
	2	Odor recognition
	3	Odor annoyance
	4	Odor intolerance
	5	Perceived intolerance
	6	Somatic irritation
	7	Toxicity

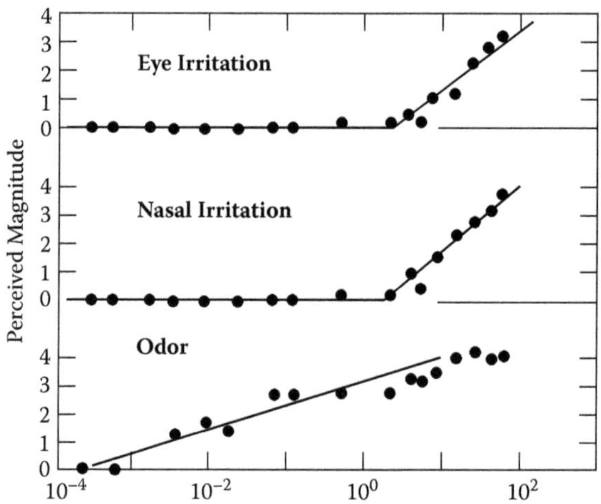

FIGURE 9.2 Relation of odor to irritation. (From Cain, W. S. in *U.S. Composting Council's 14th Annual Conference and Trade Show*, Albuquerque, NM, 2006. With permission.)

There are numerous factors that influence the detection of odors. These can be individual, environmental, and substance based. Table 9.2 provides detection thresholds for several compounds as an illustration.

Odorous Compounds

Odors are generated during the composting process. As the process of decomposition proceeds from the unstable raw feedstocks to the final stable compost products, the intensity of odors, especially malodors, diminishes. The types of odors and the compounds emitted during the composting process are a function of the feedstocks used.

TABLE 9.2
Compounds Identified During Biosolids Composting Odors

Compound	Odor Characteristic	Odor Threshold µg/m³ Low[a]	µg/m³ High	µ/g/m³ ADL[b]
	Sulfur Compounds			
Hydrogen sulfide	Rotten egg	0.7	14	6.7
Carbon oxysulfide	Pungent			
Carbon disulfide	Disagreeable, sweet	24.3	23,000	665
Dimethyl sulfide	Rotten cabbage	2.5	50.8	2.5
Dimethyl disulfide	Sulfide	0.1	346	—
Dimethyl trisulfide	Sulfide	6.2	6.2	—
Methamethiol	Sulfide, pungent	0.04	82	4.2
Ethanethiol	Sulfide, earthy	0.032	92	2.6
	Ammonia and Nitrogen-Containing Compounds			
Ammonia	Pungent, sharp	26.6	39,600	33,100
Aminomethane	Fishy, pungent	25.2	12,000	—
Dimethylamine	Fishy, amine	84.6	84.6	88.1
Trimethylamine	Fishy, pungent	0.8	0.8	0.52
3-Methylmdole (skatole)	Feces, chocolate	$4.0^{*}10^{-5}$	268	—
	Volatile Fatty Acids			
Methanoic (formic)	Biting	45.0	37,800	—
Ethanoic (acetic)	Vinegar	2,500	250,000	2,500
Propanoic (progionic)	Rancid, pungent	84.0	60,000	—
Butanoic (butyric)	Rancid	1.0	9,000	0.7
Pentanoic (valeric)	Unpleasant	2.6	2.6	—
3-Methylbutanoic (isovaleric)	Rancid cheese	52.8	52.8	—
	Ketones			
Propanone (acetone)	Sweet, minty	47,500	1,610,000	241,000
Butamone (MEK)	Sweet, acetone	737	147,000	30,000
2-Pentamone (MPK)	Sweet	28,000	45,000	—
	Other Compounds			
Benzothiozole	Penetrating	442	2,210	—
Ethanal (acetaldehyde)	Green sweet	0.2	4,140	385
Phenol	Medicinal	178	2,240	184

Source: Data from Williams, T. O. and Miller, F. C., in *Science and Engineering of Composting: Design, Environmental, Microbial, and Utilization Aspects*, ed. H. A. J. Hoitink and H. M. Keener, Ohio State University, Wooster, 1993.

[a] Low threshold indicates the lower limit of detection to most sensitive persons. High threshold value means that it is odorous to most persons.

[b] Values recalculated from volume/volume data assuming 20°C and 1 atm.

Although numerous malodorous compounds are found in the various feedstocks, the quantities do not necessarily indicate the severity of the odor. A small quantity of some compounds may be more obnoxious than other compounds present in large quantities.

There is a considerable literature on emissions from biosolids composting facilities in the United States (Van Durme et al., 1990; Wilber and Murray, 1990; Williams and Miller, 1993; Hentz, et al., 1992, Walker, 1993). Data from other feedstocks are scarce. Odor sampling and analysis are costly. Furthermore, just the presence of a compound does not necessarily indicate that a particular compound is producing an offensive odor.

Table 9.2 shows some of the odorous compounds identified during composting of sewage sludge and biosolids (Miller, 1993; Williams and Miller, 1993). The major odor groups include fatty acids, ammonia and nitrogen-containing compounds, ketones, aromatics, and inorganic and organic sulfur compounds.

Sulfur compounds and ammonia have been found in many biosolids and animal waste. Sulfur compounds are also high in food wastes, as they are a component of several amino acids. The odorous compounds most commonly found have been hydrogen sulfide, dimethyl sulfide, dimethyl disulfide, ammonia, limonene, and pinene (Hentz et al., 1992; Van Durme et al., 1990). The latter two compounds are aromatic compounds released from wood chips, which are used as a bulking agent in biosolids operating facilities. Ammonia is also often released during composting operations involving animal wastes, food wastes, and sewage sludge. Feedstocks with low carbon-to-nitrogen ratios (lower than 20:1) will release ammonia during composting. This has been a major problem with the composting of grass clippings. As the C:N ratio increases, the ammonia levels decrease. Volatile fatty acids (VFAs) can be a source of odors during the decomposition of organic matter.

Table 9.3 shows the compounds analyzed and detected from an ASP and yard waste composting facility. The compounds detected were acetone, carbon disulfide, 2-butanone (MEK), toluene, styrene, and 1,4-dichlorobenzene. One of the purposes of showing this table is to indicate that a large number of compounds analyzed were not detected.

The data on emissions are highly variable, depending on the extent and type of biosolids treatment. Early data showed many more and higher concentrations. Lees and Tockman (1987) found similar results as shown in Table 9.3. However, Van Durme et al. (1990) reported higher values for more compounds. The early data for numerous feedstocks are shown in *The Science of Composting* (Epstein, 1997).

A study published by Horowitz (2010) found that less than 5% of volatile organic compounds (VOCs) in compost emissions profiled from greenwaste operations are strong ozone precursors. Some ozone formers, like pinene and limonene, are common compounds that are emitted by trees.

Data on emissions from food waste are very limited. Table 9.4 shows early data. Studying composting of food waste with varying proportions of yard waste, E&A Environmental Consultants, Inc. (1993) found that the odorous compounds were fatty acids, mercaptans, ketones, and ammonia. Many odorous compounds are formed during anaerobic periods.

TABLE 9.3
Volatile Organic Compounds Analyzed and Detected from an ASP Biosolids and Yard Waste Composting Facility

Compound	Result µg/m³	MRL µg/m³	Result ppbV	MRL ppbV
Chloromethane	ND	33	ND	16
Vinyl chloride	ND	33	ND	13
Bromomethane	ND	33	ND	8.4
Chloroethane	ND	33	ND	12
Acetone	**3,000**	160	**1,300**	68
Trichlorofluoromethane	ND	33	ND	5.8
1,1-Dichloroethene	ND	33	ND	8.2
Methylene chloride	ND	33	ND	9.4
Trichlorotrifluoroethane	ND	33	ND	4.2
Carbon disulfide		33		10
trans-1,2-Dichloroethene	ND	33	ND	8.2
1,1-Dichloroethane	ND	33	ND	8.0
Methyl tert-butyl ether	ND	33	ND	9.0
2-Butanone (MEK)	**1,500**	33	**510**	11
cis-1,2-Dichloroethene	ND	33	ND	8.2
Chloroform	ND	33	ND	6.7
1,2-Dichloroethane	ND	33	ND	8.0
1,1,1-Trichloroethane	ND	33	ND	6.0
Benzene	ND	33	ND	10
Carbon tetrachloride	ND	33	ND	5.2
1,2-Dichloropropane	ND	33	ND	7.0
Bromodichloromethane	ND	33	ND	4.9
Trichloroethene	ND	33	ND	6.0
cis-1,3-Dichloropropene	ND	33	ND	7.2
4-Methyl-2-pentanone	**100**	33	**26**	7.9
trans-1,3-Dichloropropene	ND	33	ND	7.2
1,1,2-Trichloroethane	ND	33	ND	6.0
Toluene	**41**	33	**11**	8.6
2-Hexanone	ND	33	ND	7.9
Dibromochloromethane	ND	33	ND	3.8
1,2-Dibromocthane	ND	33	ND	4.2
Tetrachloroethene	ND		ND	4.8
Chlorobenzene	ND	33	ND	7.1
Ethylbenzene	ND	33	ND	7.5
m,p-Xylenes	ND	33	ND	7.5
Bromoform	ND	33	ND	3.1
Styrene	**49**	33	**11**	7.6
o-Xylene	ND	33	ND	7.5
1,1,2,2-Tetrachloroethane	ND	33	ND	4.7
1,3-Dichlorobenzene	ND	33	ND	5.4
1,4-Dichlorobenzene	**50**	33	**8.3**	5.4
1,2-Dichlorobenzene	ND	33	ND	5.4

Source: Courtesy of Managed Organic Recycling, Inc.

TABLE 9.4
Odorous Compounds Detected at a Composting Food Waste Facility

Material or Compound and Decomposition Condition	Product	Laboratory Analysis ppm	Field Analysis ppm
Fatty acids, anaerobic decomposition	Formic acid	2.5	ND
	Acetic acid	25.3	ND
Mercaptans and other organic sulfides, anaerobic decomposition	Mercaptans	0.32	ND
	Organic sulfides		0.2–100 in pile
Ketones, anaerobic decomposition	Methyl ethyl ketone	600	—
Ammonia, aerobic, and anaerobic decomposition	Ammonia	ND	0.05–7 in exhaust air

Source: E&A Environmental Consultants, Inc., *Food Waste Collection and Composting Demonstration Project*, City of Seattle Solid Waste Utility, Seattle, WA, 1993.
Note: ND = not detected.

TABLE 9.5
List of Malodorous Compounds Found in Fresh Manure, and Manure and Sawdust Mixture prior to Composting and after 16 Days of Composting

Compound	Fresh Manure	Manure/Sawdust Day 0 μg/g	Manure/Sawdust Day 16 Composting μg
Acetate	5,266 ± 118	6,553 ± 1,624	23 ± 1
Propionate	817 ± 27	795 ± 276	ND
Butyrate	385 ± 42	ND	ND
Valerate	70 ± 11	ND	ND
Isobutyrate	78 ± 4	77 ± 35	ND
Isovalerate	127 ± 7	51 ± 18	ND
Phenol	35 ± 3	19 ± 2	ND
p-Cresol (*o*- and *m*-)	197 ± 4	17 ± 4	ND
Indole	5 ± 2	ND	ND
Skatole	0.8 ± 0.8	TR	ND

Source: Elwell, D. L. et al., *Compost Sci. Util.*, 12, 102–7, 2004. With permission.
Note: ND = none detected, TR = trace.

Effect of Composting on VOC Destruction

Elwell et al. (2004) studied the effects of composting on odorous compounds emitted by dairy manure (Table 9.5). The study involved both fresh manure and manure with sawdust. Composting was very effective in elimination of malodorous compounds.

Rosenfeld et al. (2004) measured the concentration of odors during biosolids composting from a windrow, static pile, and biofilter. Some of the data are shown in Table 9.6. Emissions above a windrow were more than three times than those over a static pile. The biofilter captured over 98% of the odor emissions.

TABLE 9.6
Odorant Concentrations in Air Samples Collected above the Windrow and Aerated Pile before and after Biofilters

Odorant	Sample Location	Odorant Concentration μg/m³	Aeration Pile Odor Reduction Compared to Windrow
Ammonia	Above windrow	239,483	
	Above aerated pile	66,492	72%
	Before biofilter	98,442	
	After biofilter	1,658	
Dimethyl disulfide	Above windrow	<192	ND
	Above aerated pile	<192	
	Before biofilter	11,081	
	After biofilter	961	
Carbon disulfide	Above windrow	<155	ND
	Above aerated pile	<155	
	Before biofilter	1,951	
	After biofilter	1,305	
Formic acid	Above windrow	3,650	
	Above aerated pile	1,583	57%
	Before biofilter	1,675	
	After biofilter	<60	
Acetic acid	Above windrow	12,100	11%
	Above aerated pile	10,767	
	Before biofilter	9,950	
	After biofilter	6,600	
Sulfur dioxide or carbonyl sulfide[a]	Above windrow	<131	ND
	Above aerated pile	<131	
	Before biofilter	1,441	
	After biofilter	<131	

Source: Rosenfeld, P. et al., *Water Environ. Res.*, 76, 310–15, 2004. With permission.
[a] The peaks for sulfur dioxide and carbonyl sulfide overlapped and the total peak area was used.
Note: ND = nondetactable in air samples above the windrow or static pile.

Data from MSW facilities are meager. Kim et al. (1995) measured the amount of volatile organic compounds (VOCs) at an MSW composting facility. They found that chloroform, toluene, methylene chloride, ethylbenzene, and trichloroethylene were detected in descending order of concentration.

Peterson et al. (2000) characterized the emissions from two yard waste composting facilities. That data are shown in Table 9.7. They found five compounds: 2-butanone, 4-methyl-2-pentanone, xylenes, carbon disulfide, and ammonia. In some cases, the concentrations of these compounds downwind from the facilities were quite similar to upwind or background concentrations. Ammonia exceeded the USEPA risk-based

TABLE 9.7
Compounds Detected in Air at Two Yard Waste Composting Facilities

Compound	Upwind μg/m³	Downwind μg/m³
	Volatile Organic Compounds	
Acetone	15.5–26	12.2–23
Benzene	1.3–2.4	1.2–1.4
2-Butanone	2.9–3.6	4.3–4.9
Ethylbenzene	ND–TR	ND
4-Methyl-2-pentanone	ND	1.2–1.7
Toluene	3.1–9.7	2.1–3.9
m,p-Xylene	3.5–1.0	1.6–1.5
o-Xylene	ND–1.3	ND
Vinyl acetate	ND–1.4	ND–0.92
	Halogen Compounds	
Carbon tetrachloride	TR–0.8	TR–0.75
Chloromethane	1.8–2.2	ND–2.0
1,1-Dichloroethene	ND–5.0	ND
Methylene chlorine	ND–1.7	ND–1.2
Tetrachloroethene	ND–10.9	ND–1.4
1,1,1-Trichloroethane	TR–1.3	ND
Trichloroethene	ND–2.7	ND
Trichlorofluoromethane	1.7–1.5	1.5–1.6
Trichlorotrifluoroethane	1.3	1.2
	Reduced Sulfur Compounds	
Carbon disulfide	ND–1.7	1.8–4.3
	Nitrogen Compounds	
Ammonia	ND	36–132

Source: After Peterson, M. K. et al., in *Odors and VOC Emissions 2000*, Water Environment Federation, Cincinnati, OH, 2000.

ND = not detected; TR = trace

concentrations (RBCs) and the odor threshold. However, the authors did not include amines or carbonic acid esters. The authors concluded that the presence of the compounds identified does not pose a health risk to residents nearby. They based this on the fact that ammonia dissipates very rapidly and that residents were at a distance from the facility.

A study by Heida et al. (1995) measured the indoor air of a closed facility's composting garden refuse and organic household wastes. They found aromatic hydrocarbons, aliphatic hydrocarbons, limonene, and hydrogen sulfide. Limonene, a compound emitted from wood wastes, was at the highest concentration, followed by aliphatic hydrocarbons and alkylbenzenes. They attributed the high concentration of alkylbenzenes and aliphatic hydrocarbons to be caused by limonene, a natural compound in wood.

A study was conducted to measure emission rates from windrows in lb/day from greenwaste with and without food waste addition (CIWMB, 2007). After fourteen days, the emission rates dropped significantly. Thus, a major control strategy is to reduce the emissions during this early period (see Chapter 10). The total emission rate for the food waste over fifty-seven days was 37.6 kg/day (82.9 lb/day), with 94% occurring during the first fourteen days. Emissions from greenwaste were considerably lower, and after fifty-seven days were 15.4 kg/day (34 lb/day). During the first three days, over 60% of the emissions occurred. In the study, the addition of 15% food waste to greenwaste resulted in nearly twice the VOC emissions over the fifty-seven days (Figure 9.3).

The design of facilities and their management can greatly reduce VOC and odor emissions, as discussed in Chapter 10.

Table 9.8 shows data from an MSW aerated-turned-windrow composting facility. Due to the heterogeneous nature of MSW, considerably more compounds were

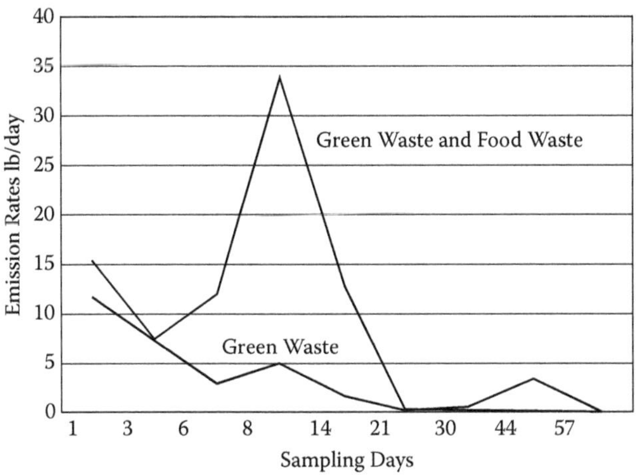

FIGURE 9.3 Emission rates from food waste and greenwaste windrows. (From CIWMB, *Emissions Testing of Volatile Organic Compounds from Greenwaste Composting at the Modesto Compost Facility in the San Joaquin Valley*, California Integrated Waste Management Board, Sacramento, 2007. With permission from CalRecycle.)

TABLE 9.8
Surface VOC Emissions from a Solid Waste (MSW) Windrow Operation

Compound	Process Day 8 µg/m³	Process Day 16 µg/m³	Process Day 43 µg/m³
Acetone	1,215	560	117
Vinyl acetate	920	198	17
2-Butanone	8,750	1,255	21
Toluene	43	6	5.5
Tetrachloroethene	13	0	0
Ethylbenzene	36.5	5.5	5
Styrene	179	51.5	24
m,p-Xylene	170	40.5	5.9
o-Xylene	96	25.5	3.4
Acetic acid	150	0	0
Ethanol	1,500	250	0
Isopropanol	150	0	0
Methyl acetate	250	100	0
2-Methyl propanol	150	0	0
1-Propanol	100	0	0
2-Butanol	1,950	0	0
3-Methyl Butanol	600	55	0
C5-Alcohol	200	0	70
Hexanal	100	250	70
C8H18 Alkane	100	65	0
Alpha pinene	3,000	990	55
Terpenes	7,950	1,000	0
Beta pinene	2,500	800	65
d-Limonene	15,500	7,500	800
C11H24 alkane	3,000	1,150	0
C11 H22 hydrocarbon	1,500	0	0
n-Undecane	3,500	1,250	0
Camphor	1,500	0	25
C12H26 alkane	950	0	0

detected. It is very evident that many of these compounds are either volatilized or degraded. At days 16 and 43 the concentration of most of these compounds dropped significantly. Many compounds, such as pinene and limonene, are a product of wood and are usually found in high concentrations.

There are very little data on VOC emissions from food waste as related to technologies. Nothing has been published in referred journals. In 2008, an air emissions data review was made for San Joaquin Valley Air Pollution Control District of data generated by the California Integrated Waste Management Board. These data are presented in Figures 9.4 and 9.5. Figure 9.4 is for volatile organic carbon compounds and Figure 9.5 is for ammonia. In Figure 9.4 peak and average VOC emissions were

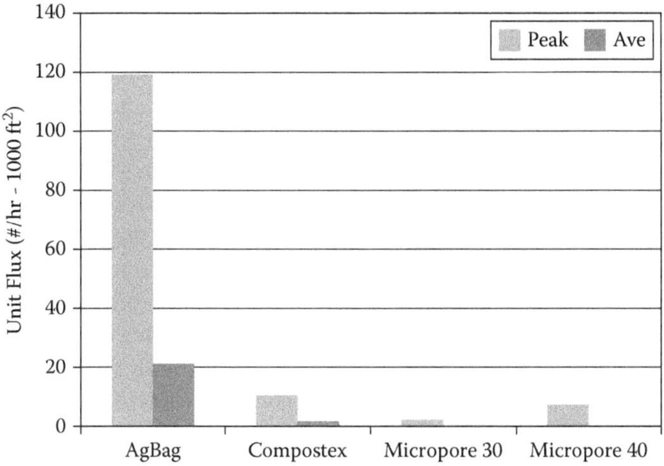

FIGURE 9.4 VOC emissions in lb/h-1,000 ft^2 from three different technologies. (From Card, T. R. and Schmidt, C. E., *Organic Material Composting and Drying Focusing on Greenwaste Compost Air Emissions Data Review*, San Joaquin Valley Air Pollution Control District, CA, 2008.)

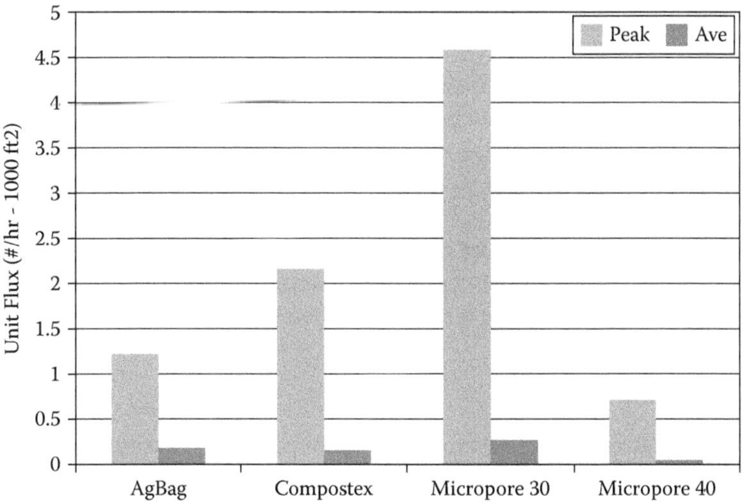

FIGURE 9.5 Ammonia emissions in lb/h-1,000 ft^2 from three different technologies. (From Card, T. R. and Schmidt, C. E., *Organic Material Composting and Drying Focusing on Greenwaste Compost Air Emissions Data Review*, San Joaquin Valley Air Pollution Control District, CA, 2008.)

greatest from the Ag-Bag and least from the two micropore covers. Micropore 30 had the lowest VOC emissions from composting food waste.

Peak and average ammonia emissions were greatest with micropore 30. Micropore 40 had the least ammonia emissions. The micropore covers are made from expanded polytetrafluoroethelene (PTFE) membranes. They are designed to maximize oxygen transfer while minimizing water evaporation. Their pore size is a barrier to most non-methane hydrocarbons, but not generally to ammonia (Card and Schmidt, 2008).

From all the data presented here, it is clear that during the first two weeks most of the emissions occur, and odor potential is highest. Therefore, it is imperative to devise techniques to reduce odors and VOC emissions during the first fourteen days. Furthermore, once regulatory requirements for pathogen and vector attraction are accomplished, the temperature could be reduced from thermophilic to mesophilic. At mesophilic temperatures, considerably less VOCs and odors would be produced.

Relation of Odors to VOC Emissions

There is no direct relationship of VOC emissions to odor production. Some VOC compounds at very high concentrations, such as acetone, do not produce malodors. Others, such as limonene, have an odor that is not necessarily offensive. Still others, such as disulfides, at can be very offensive at very low concentrations. Therefore, a panel using a standard method measures odors (described in a separate section). There is also a difference between the U.S. and European standards. This sometimes becomes a problem when analyzing data from other countries. VOCs are analyzed by chemical analysis using gas chromatography, mass spectrometry, or other techniques. VOC analysis can at times be used to identify the odor source, or even indicate that the composting facility was not the major source of the odor. The source may be an industrial or other type of agricultural activity.

Odors are essentially a local and personal issue, whereas VOCs are an environmental issue. A person living near a bakery or a chocolate factory smelling what normally is considered to be pleasant odors may complain of the odor to which he is exposed daily at high concentrations. Not only do VOCs affect local air quality, but also some are greenhouse gases.

In California, odors are under the jurisdiction of the California Integrated Waste Management Board (CIWMB). Air quality boards have jurisdiction of VOCs. This sometimes causes conflict with respect to regulatory actions.

Odor Characterization

Odors are characterized in several ways as to quantity, intensity, persistence, and characteristics.

Odor Quantity

Odor levels are usually expressed as a dilution-to-threshold ratio (D/T) rather than a concentration. D/T values are determined by an odor panel of eight to ten people and express the number of dilutions required by 50% of the panel to still detect the odor

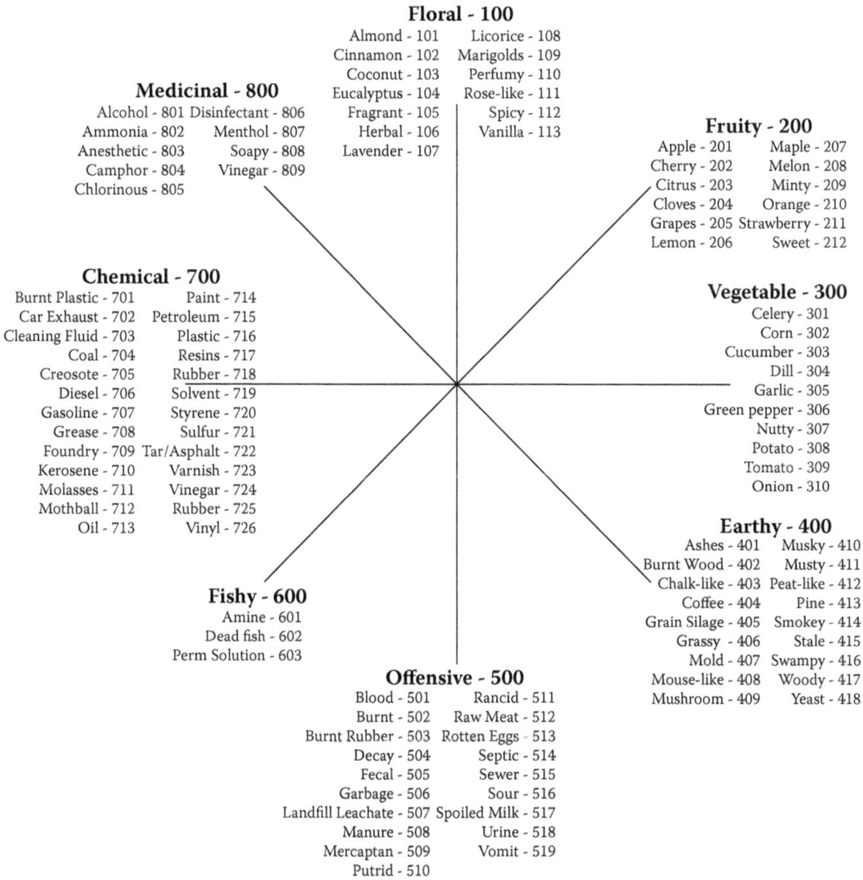

FIGURE 9.6 Qualifying characteristics of odors. (Courtesy of St. Croix Sensory, Inc.)

(ED_{50}). One of the major reasons for not characterizing the odor by concentration of specific compounds is that one compound can be present in a large concentration, and yet the malodor effect is relatively low, whereas another compound is extremely offensive even at a very low concentration. The quality description of odors is shown in Figure 9.6. Eight odor descriptors are used. It should be remembered that the quantity D/T is very broad. The D/T value near a bakery or chocolate factory could be very high, but would not be offensive as the odor having less D/T values near a wastewater treatment plant or landfill.

Odor Intensity

This is the relative strength of the odor compared to concentrations of a standard compound, usually *n*-butanol. It is expressed in parts per million (ppm) of butanol. A large value of butanol indicates a stronger odor, whereas a small concentration of butanol indicates a weaker odor. The relationship between intensity and concentration is expressed by the equation (WEF, 2004)

$$I = kC^n$$

where:

I = intensity
C = mass concentration of the odorant in mg/m^3
k and n = constants that are different for every specific odorant or mixtures of specific odorants

Odor intensity of the ambient air can be measured using ASTM method E544-99, *Standard Practice for Suprathreshold Odor Intensity Measurement* (ASTM, 1999). Once the intensity (butanol equivalent concentration) of an odorant is known, its total mass per unit time can be used to estimate the mass of odor by multiplying it times the volume of odorous air per unit time (Walker, 1993).

Odor Persistence

Odor persistency is an indication of how long the odor remains and is an indication of the rate of dilution. Odor persistence can be quantified and represented as an odor response function (McGinley and McGinley, 1999b). The pervasiveness is determined by progressively diluting the odor and measuring the intensity at each dilution. Figure 9.7 illustrates a plot of odor intensity vs. dilution of pervasive and lesser pervasive odors. The flatter the slope, the more pervasive the odor (Walker, 1993).

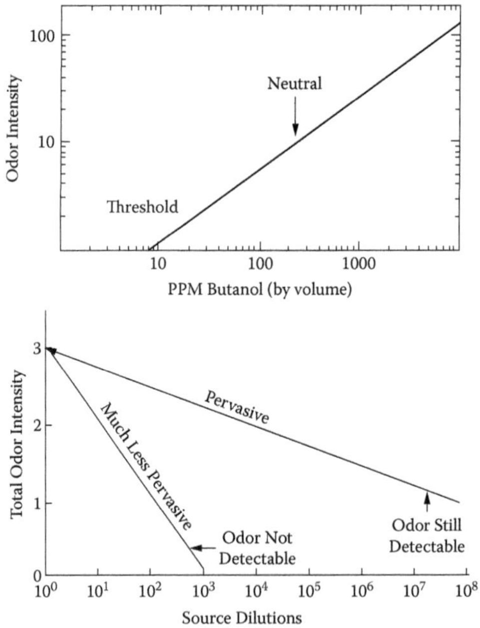

FIGURE 9.7 Odor characterizations as related to total odor intensity for pervasive and less pervasive odors. (Data from Walker, J. M. in *Science and Engineering of Composting: Design, Environmental, Microbiological, and Utilization Aspects*, ed. Hoitink and Keener, Renaissance Publications, Worthington, OH, 1993.)

Odor Character

Character denotes the type of odor (chemical makeup) or its offensiveness. Odor character or quality is the property that differentiates it from another odor. Odors are classified based on descriptive terms. For example, hydrogen sulfide smells like rotten eggs, and dimethyl sulfide smells like rotten cabbage. Odor character is evaluated by comparison with other odors, whether directly or with descriptors, which is an important factor when identifying odor nuisance violations.

Hedonic Tone

The hedonic tone is the relative pleasantness or unpleasantness of the odor. There are many different scales to rank the hedonic tone. One relatively simple scale is B10 to +10, where 0 is neutral, B10 is extremely unpleasant, and +10 is extremely pleasant. Odors with a strong negative hedonic tone (less than 0) will have lower nuisance thresholds than those with a more neutral or positive tone (0 to +10). There is a difference between the acceptability and the hedonic tone of an odor. Acceptability is subjective and depends on the person experiencing the odor. For example, an otherwise pleasant odor may be unacceptable if it persists as part of an air pollution problem in a residential area.

An individual's experience and emotional associations may also determine the degree of pleasantness and unpleasantness of an odor.

Odor Measurements

The measurement of odors or odorous compounds can involve field or laboratory measurements. Field measurements can be used either for specific compounds or for general odor intensity. Measuring specific compounds can be achieved by (1) collection of a sample at a point source or area source and analyzing it in a laboratory, (2) using detection tubes, or (3) using instruments such as a Jerome meter for measuring H_sS, a flame ionization detector, or an organic vapor analyzer. General odor intensity can be measured in the field by olfactometry (AWMA, 2002).

Collecting emissions from a point or area source could be done over a pile or windrow, inside a building at a specific location, an exhaust, or at some location within the facility that is believed to be a source of odors. Samples are usually collected in a Tedlar® bag or Summa canister and a sample is sent to a laboratory for analysis for specific compounds. The collection of a sample is usually done with an isolation emission flux chamber developed by USEPA as shown in Figure 9.8 (USEPA, 1989). A modification of the sampling equipment and technique used by the California Integrated Waste Management Board is shown in Figures 9.9 and 9.10. Figure 9.11 shows the collection of odors and VOCs in the field.

Often, and more effective from the point of odor management and control, odors are measured as a whole rather than a specific compound. This is done by either using a field olfactometer (Figure 9.12) or collecting air samples in Tedlar bags and submitting them to a laboratory for analysis by a panel. Odor panels, usually eight to ten people, are selected based on *Guidelines for the Selection and Training of Sensory Panel Members* (ASTM 1981). Gas samples are diluted to below detection, and then

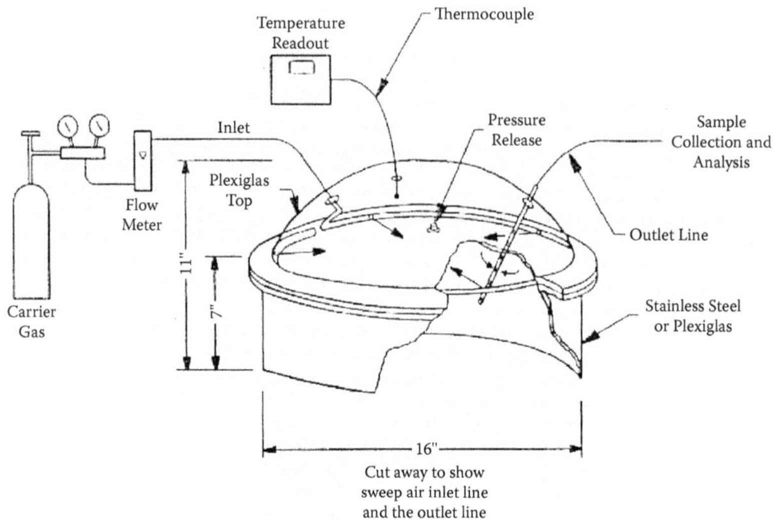

FIGURE 9.8 Surface isolation emissions flux chamber used to collect odorous compounds for analysis from unaerated or negatively aerated piles as suggested by the USEPA.

concentrations are increased until 50% of the panel can detect the odor. The concentrations are expressed as the number of dilutions that were required by 50% of the panel to detect the odor and recorded as dilution to threshold (D/T). For example, a D/T of 5 means that four volumes of odor-free air must be mixed with one volume of the odorous air (for a total of five volumes of air) to dilute the sample to the required concentration (Mahin, 1998; McGinley and McGinley, 1999a; EN 13725, 2003).

Odor Modeling

Odor dispersion modeling is a technique for estimating odor concentration caused by emissions from a source. While it is not possible to fully model the complexities of the atmosphere and the exact transport and dispersion of an odor, a series of mathematical formulas have been developed from empirical and theoretical studies to reliably estimate odor concentrations. Incorporating these formulas in two computer-based models greatly increases the ability to model numerous sources and receptors, as well as quickly and efficiently compare various odor control strategies and their impact on odor concentrations at receptor locations. In particular, dispersion models allow one to readily evaluate the changes in concentrations at receptors due to changes in the size of facility, modifications of operations, size or concentrations of odor sources, and the climatologic changes. They can be used to evaluate the impact of different systems and configurations and predict worse-case scenarios.

When planning for new facilities, odor modeling can be a very useful tool in predicting the transport and potential impact of odors to receptors. Odor modeling can also be a useful tool to determine modifications of operations, or the implementation of design changes, and how they may affect odor emissions and impacts to receptors.

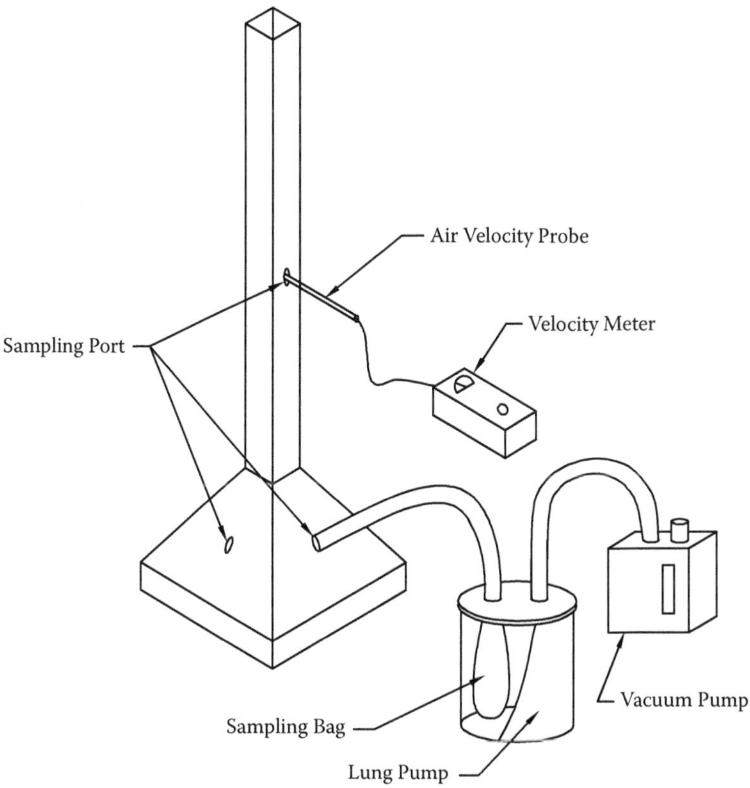

FIGURE 9.9 A modification of the sampling equipment and technique used by the California Integrated Waste Management Board. (From CIWMB, *Emissions Testing of Volatile Organic Compounds from Greenwaste Composting at the Modesto Compost Facility in the San Joaquin Valley*, California Integrated Waste Management Board, Sacramento, 2007.)

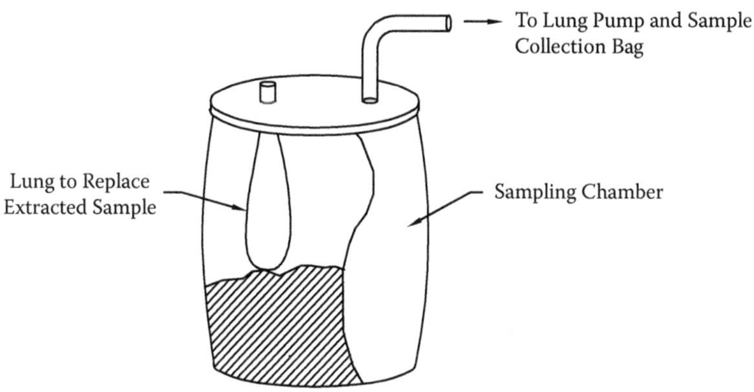

FIGURE 9.10 Cold trap used to collect VOCs. (From CIWMB, *Emissions Testing of Volatile Organic Compounds from Greenwaste Composting at the Modesto Compost Facility in the San Joaquin Valley*, California Integrated Waste Management Board, Sacramento, 2007.)

FIGURE 9.11 Using the surface emissions flux chamber to measure odor emissions in the field.

FIGURE 9.12 Using an olfactometer to measure and identify odor sources in the field. (Courtesy of St. Croix Sensory, Inc.)

The computer model generally is used to predict the maximum odor concentrations and number of odor impacts. In the past the model ISCST3 was used. The AERMOD modeling system officially replaced ISCST3 in December 2006 as the preferred and approved regulatory model for simulating the impacts of emissions from a variety of sources, including power plants, industrial facilities, landfills, hazardous waste facilities, and composting facilities (Diosey, 2008). Unlike ISCST3, the AERMOD modeling system consists of three programs. The AERMOD meteorological preprocessor (AERMET) develops the meteorological input file from surface and upper air data, while the AERMOD terrain preprocessor (AERMAP) generates terrain and receptor grid input files from digitized terrain data. The AERMOD dispersion model then uses source information, combined with the AERMET and AERMAP output, to determine the concentrations at the specified receptor locations. Because AERMOD has the ability to characterize the profile of the planetary boundary layer (PBL) and the dispersion of pollutants within it more realistically, the model requires a significantly greater amount of site-specific input for the user to evaluate, and more options to select (Diosey, 2008). The ISCST3 model uses local topographical and meteorological data, as well as specific site inputs, to generate odor dispersion patterns. U.S. Geological Survey (USGS) maps from a location and meteorological data collected from the best meteorological station are used in the model. The USEPA recommends the use of five years of meteorological data. All of the available meteorological data covering a wide range of atmospheric conditions are used. Odor emissions data from comparable operating facilities need to be used for each of the potential odor sources. Model inputs are listed below:

- Spatial locator grid
- Potential odor receptors
- Topography data
- Climatologic data (historic)
    - Temperature
    - Wind speed
    - Wind direction
    - Stability class
- Odor emission data
    - Size of source
    - Strength of source
    - Type of activity
    - Duration of activity

Meteorological data are most commonly obtained from the National Climatic Data Center (NCDC), which maintains records of meteorological conditions recorded at weather stations and airports throughout the country. Odor models can accept more than four hundred receptor locations placed on a grid system. Discrete receptors (i.e., specific residences) can also be entered into the model. The model estimates the concentration at each receptor for each hour of meteorological data considered (typically a full year, or 8,760 hours). The model produces summaries of the highest concentrations recorded at each receptor during that year, as well as extensive files that record for each receptor each occurrence of a concentration higher than a user-defined threshold. Listed below are the potential model outputs:

- Maximum levels at each receptor
- Isopleths of concentrations
- Frequency of nuisance odor conditions
- Weather conditions when levels are highest

AEROMOD and ISCST3 utilize hourly meteorological data to define the conditions of the plume contributing to pollutant dispersion. The models estimate pollutant concentrations for each source and receptor for each hour of input meteorological data, and calculate a user-specified, short-term average concentration. For example, if the selected short-term average is two hours, the concentration is calculated at a receptor for each hour of meteorological data. The two resulting concentrations from two hours of meteorological data are averaged for the resulting concentration (mean concentration). Humans can detect odor at much shorter durations than one hour; therefore, an averaging time of less than one hour must be determined to calculate the short-duration concentration (peak concentration).

Regulated averaging times used for determining the short duration or peak concentration for odors range from five minutes to one hour based on existing state odor regulations. Exposure to an odor for five minutes would not represent a nuisance, and such regulations would contribute to overly conservative peak odor concentrations. Based on odor analysis experience, ten minutes is the minimum exposure time resulting in a nuisance.

The model results are plotted as isopleth lines, indicating the maximum ten-minute odor concentration to be expected in the area surrounding the planned location where a composting facility is to be located or where one exists. In addition, specific receptor locations are selected from USGS aerial photographs showing structures near the facility to provide a maximum ten-minute odor concentration and number of impacts predicted to occur at the nearest receptors.

Once an odor peak averaging time has been selected, a conversion factor is applied to the model to convert from an hourly concentration (mean concentrations) to a concentration of less than one hour (short-duration concentration or peak concentration). Several studies have been conducted to determine the relationship between the peak concentration and mean concentration (peak-to-mean ratio). Hino (1968) suggested using the –1/2 power law for sampling times from ten minutes to five hours and recommended using the –1/5 power law for sampling times of less than ten minutes.

Using the –1/5 power law, the peak-to-mean concentration ratio can be calculated as follows:

$$C_2/C_1 = (t^2/t^1)^{-1/5} \quad (9.1)$$

where:

C_1 = concentration estimate for time t_1
C_2 = concentration estimate for time t_2
t_1 = mean averaging time
t_2 = peak averaging time

Using Equation 9.1, a ten-minute to one-hour peak would mean the concentration ratio equals 1.43. This value was used in the modeling input for the model runs performed.

Mahin (1998) indicated that one of the more commonly accepted conversions from one averaging time to a shorter averaging time is

$$X_s = X_{model}\,(T_{model}/T_s)^p$$

where:

- X_s = D/T level at a shorter averaging time
- X_{model} = D/T level at averaging time predicted by model
- T_{model} = model predicted averaging time
- T_s = shorter averaging time
- P = power law exponent usually used as 0.2

Examples of the result of odor modeling on the potential impact to receptors are shown in Figures 9.13 and 9.14. In Figure 9.13 receptors near the composting facility, where the D/T are greater than 5, could experience odor problems. Conducting the model to enclose the biofilter reduced the potential of odors to a D/T of less than 1. Typically an emission rate of 25 D/T is assigned to a biofilter, even though the odors are not malodorous and are more of a woody, compost type. However, enclosing the biofilter and further treating the odors can greatly reduce emissions (Figure 9.14).

Odor concentrations resulting from previous sampling at existing biosolids, food waste, and yard waste composting facilities are used to determine odor emission rates for each odor source, including feedstock tipping, feedstock storage, feedstock mixing, feedstock transfer, grinding, composting technology, teardown of piles or

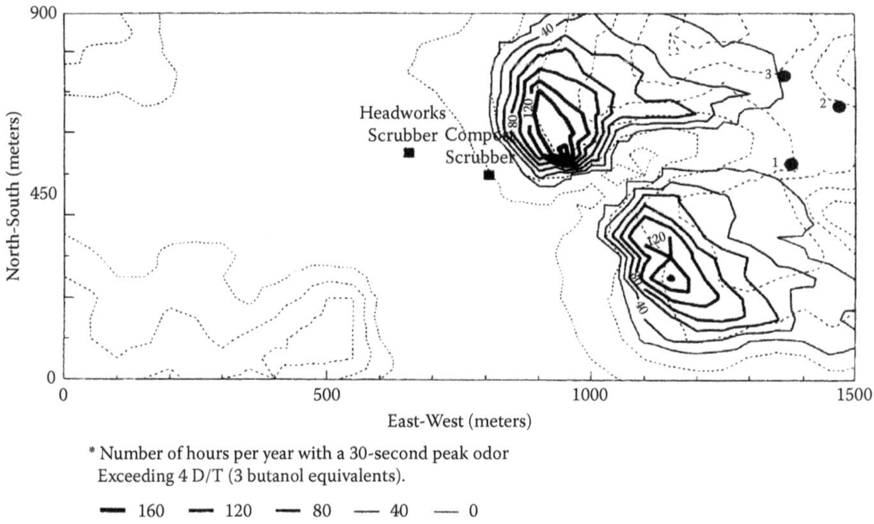

FIGURE 9.13 Results of an odor model for a proposed ASP composting facility using negative aeration and an open biofilter.

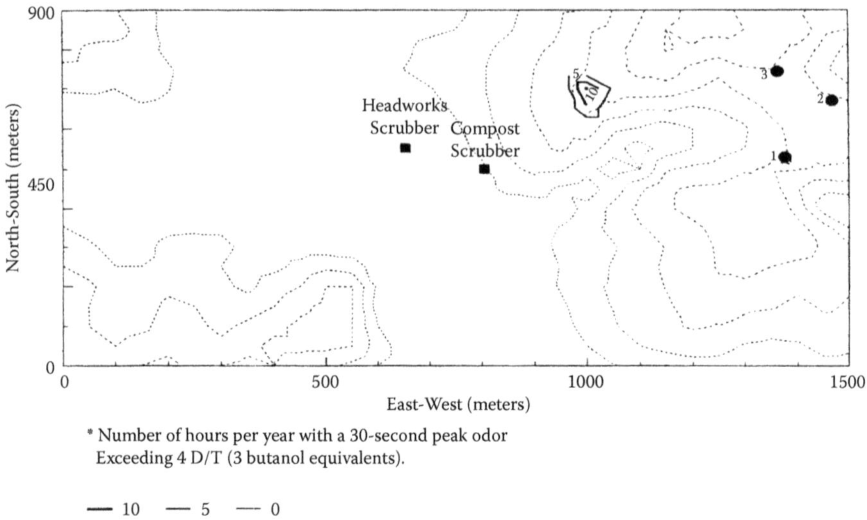

FIGURE 9.14 Odor model for an enclosed composting facility and enclosing the biofilter.

windrows, screening, product storage, and product load-out. The emission rates for each of the odor sources are then calculated.

PRODUCTION OF ODORS AS RELATED TO TECHNOLOGY

The amount of odor generated during the composting period depends on the composting technology employed.

Table 9.9 shows the sources and magnitude of odors from a windrow facility. In Table 9.9 over 67% of the odors occurred within twenty-seven days of windrow composting. Biosolids storage was another 15%. The control of the biosolids odors from storage is easy and could be done by capturing the emissions and treating them in a biofilter. In contrast, in Table 9.10 over 84% of the odors occurred during the first three weeks of windrow composting. However, the intensity of odors is greatest immediately after turning. Consequently, in a large facility where turning is continuous, there is a high emission rate. Brenk and Garibay (1995) reported that emissions of ammonia and methane were greatest during the first twenty-two days of the compost cycle. Controlling odors from windrows is difficult and is discussed in Chapter 10.

Table 9.11 shows the sources and magnitude of odors from an aerated static pile (ASP) system. Ninety-five percent of the odors were produced during composting, with another 3% in curing. In Table 9.13, 75% of the odors were attributed to composting and an additional 10% to curing. Three times as much odor was produced during windrow composting than with the ASP system. Several studies have shown that odor emissions are greatest during the first seven to ten days. Consequently, it is recommended that aeration in an ASP be in the negative mode for the first ten to fourteen days and then be reversed.

TABLE 9.9
Emissions Data and Relative Contribution for a Windrow Composting Facility

Odor Sources	Odor Concentration (D/T)	Odor Generation Rate (ou/s-m²)	Total Contributing Area (m²)	Odor Emission (ou/s)	Relative Odor Contribution (%)
		Area Sources			
Yard waste storage	106[a]	0.68	1,600	1,100	0.53
Recycle storage	126[a]	0.81	1,300	1,100	0.51
Food waste storage	2,620[b]	17	240	4,100	1.98
Biosolids storage	57,974[a]	370	83	31,000	15
Windrows 0–6 days old	1,252[c]	8.0	6,900	56,000	27
Windrows 7–11 days old	429[c]	2.7	6,300	17,000	8.6
Windrows 12–27 days old	586[c]	3.8	17,000	65,000	32
Windrows 28–61 days old	144[c]	0.92	22,000	23,000	11
Windrows 62–90 days old	58[c]	0.37	16,000	5,800	2.9
		Volume Sources			
Grinding operations	6,683[d]			1,500	0.71
Feedstock tipping	200[e]			29	0.014
Feedstock mixing	1,500[e]			220	0.11
Feedstock transfer	82[e]			12	0.006
Compost windrow building	82[e]			12	0.006
Compost windrow turning	586[c]			85	0.04
Compost windrow teardown	82[e]			6.4	0.003
Curing windrow building	82[e]			6.4	0.003
Curing windrow turning	157[c]			12	0.01
Curing windrow teardown	89[c]			6.6	0.003
Screening	1,000[e]			149	0.07
Product load-out	1,000[e]			66	0.03
Total inputs				**203,000**	**100%**

[a] Data collected at Columbus, Ohio, biosolids compost facility.
[b] Data collected at Yuma, Arizona, municipal solid waste pilot study aerated windrow, day 1.
[c] Data collected at Newby Island, California, yard waste compost facility.
[d] Data collected at Amherst, New York, yard waste compost facility.
[e] Data collected at Northwest Cascade, Washington, biosolids, food waste, and yard waste compost facility.

TABLE 9.10
Sources and Magnitude of Odors during Windrow Operations

Source	Odor Flux Ou/s-m²	Surface Area m²	Duration h	Duration s	Total Odor Units (10⁶)	Percentage
1-week-old rows	1.3670	1,380	24	86,400	163.00	42.4
3-week-old rows	1.3608	1,380	24	86,400	162.25	42.2
5-week-old rows	0.1740	1,380	24	86,400	20.75	5.4
7-week-old rows	0.0311	1,380	24	86,400	3.71	1.0
Curing piles	0.1239	1,395	24	86,400	14.93	3.9
Storage	0.6285	149	24	86,400	8.09	2.1
Turning	989.6	1	1	3,600	3.56	0.9
Mixing	0.7765	1	1	3,600	0.002	0
Pile breakdown	2,332.8	1	1	3,600	8.40	2.2
Screening	38.556	1	1	3,600	0.14	0
Total					384.83	

TABLE 9.11
Odor Emissions during an ASP Operation

Source	Odor Concentration (D/T)	Duration (h/day)	Total Emissions (10³ odor units/ day)	Percentage
Feedstock delivery	200	8	77	<0.0
Feedstock mixing	1,500	8	468	<0.0
Composting				
Pile construction	82	8	26	
Surface (1–5 days)	648	24	403,685	72.3
Surface (6–14 days)	449	24	111,886	20.0
Surface (15–28 days)	82	24	14,303	3
Pile teardown	82	8	26	
Curing				3.3
Surface (1–7 days)	82	24	7,802	1
Surface (8–28 days)	82	24	2,601	
Surface (29–70 days)	82	24	7,802	1
Pile teardown	82	8	26	
Postprocessing				1.7
Screening	82	8	26	
Storage	82	24	9,493	1.7
Transfer	82	8	26	
Total			558,243	

Table 9.11 provides an example of odor emissions during an ASP operation. Composting contributed to 94.9% of odor emissions. During the first five days, 72% of the odors were emitted, and another 20% during the six- to fourteen-day period. Odors from ASP operations are more easily controlled (see Chapter 10). Similar data are shown in Table 9.12.

The data depicted in the above-mentioned tables can vary as a result of the technology, facility design, and management. However, these data are valuable to both regulators and operators. They can indicate the most likely sources of odors and what management changes may be most effective.

It is much easier to contain and manage odors from ASP operations than from windrows. This is discussed in Chapter 10. Selection of the proper composting method often depends on site conditions and proximity to sensitive receptors. This could determine whether a facility can operate in the open or needs to be partially or completely enclosed. The two most common composting technologies used are either static systems or agitated systems. Odors can emanate from either a point source or an area source. A point source can be the exhaust from a blower. An area source can be a windrow surface, biofilter surface, or a pile of compost being cured. The behavior and transport of odors are related to meteorology and topography. Generally, point source odors are more concentrated than area sources. Therefore, point sources are easier to control.

Walker (1993) points out that under stable meteorological conditions, i.e., vertical mixing is almost nonexistent, as in the case of inversions, the plume can travel large distances without spreading, diluting, or reaching the ground. The odor plume would not cause a problem unless elevated topography would cause it to touch down where there were human odor receptors.

In the case of an area source during relatively stable conditions, i.e., little mixing, odors can travel virtually undiluted along the ground surfaces. These undiluted odors can reach receptors nearby and result in complaints. The severity of these complaints would depend upon the quantity, intensity, and pervasiveness of the odor, as well as the sensitivity of the human odor receptors. Odors from a point source such as a building can be captured and treated in a biofilter or other odor management technologies.

TABLE 9.12
Sources and Magnitude of Odors during Aerated Static Pile Composting

Source	Odor Flux Ou/s-m²	Surface Area m²	Duration h	Duration s	Total Odor Units 10^6	Percentage
Aerated pile	0.314	3,323	24	86,400	90.15	75.1
Aerated cure	0.0318	4,506	24	86,400	12.38	10.3
Biofilter	0.508	130	24	86,400	5.71	4.8
Mixing	0.7765	1	1	3,600	0.002	0
Pile breakdown	989.6	1	1	3,600	3.56	3.0
Screening	38.56	1	1	3,600	0.139	0.1
Storage	0.6285	149	24	86,400	8.09	6.7

ODORS AND RECEPTORS

FACTORS AFFECTING THE ODOR INFLUENCE SENSITIVE RECEPTORS

The main factors affecting the dispersion of odors and their potential impact on sensitive receptors are climate, microclimate, and topography.

Odors are carried away from the source by wind and diluted by mixing with the ambient air. If odors move off-site before they are diluted, odor impacts are more likely to occur. Prevailing wind direction, wind speed, and atmospheric inversions can greatly affect odor impact to receptors. Atmospheric inversions may limit the vertical mixing of air. Inversions affect air quality when the warm air layer above the cool air on the earth's surface acts to prevent mixing, which disperses pollutants. Surface inversions can take place on clear, cool nights, when the earth's surface radiates heat away rapidly. If the air is clear, the ground and the air directly above it can be cooler than the air at higher altitudes. Surface inversions can also occur at night in valleys, when cold, dense air flows downslope under the influence of gravity, draining off the slopes and uplands, and into the valleys, resulting in colder air in the bottom of the valley and warmer air above. Odors can become trapped in the inversion layer. When the temperature increases, especially early in the morning, odors can move off-site and impact receptors. Microclimate is the climate affected by local conditions. This can be the result of fog and dense air in swales, low areas, near bogs, etc.

Topography is an important factor affecting odor concentrations on sensitive receptors. Generally, odor sources located at elevations higher than sensitive receptors are less apt to affect receptors because odors tend to mix and disperse in the air layers above the receptors. Under certain meteorological conditions, odors generated at higher elevations can be carried into lower elevations. This could occur when winds sweep down a mountain or a ridge into a lower area.

ODOR AS A HEALTH ISSUE

Environmental odor incidences are a significant portion of the air pollution complaints received by USEPA and state regulatory agencies. Odors from wastewater and large animal operations have resulted in health complaints such as eye, nose, and throat irritations, headache, diarrhea, hoarseness, sore throat, cough, chest tightness, nasal congestion, palpitations, shortness of breath, stress, drowsiness, and mood alterations (Schiffman and Williams, 2005).

Odors emitted from composting operations are often reported as a nuisance. Knasko (1993) evaluated the effects of intermittent bursts of pleasant, unpleasant, and no odor on human task performance, mood, and perceived health. They reported that odors did not influence any of these measures. However, subjects exposed to malodors subsequently felt that odors had a negative effect on all these conditions. Cain (1987) indicated that people associated the health or harmful effects of environments by the presence or quality of perceived odors. Dalton (1999) reported that many health-related effects of exposure to odorants are mediated not by direct effect of odors, but by cognitive association of odors and health. In a study, Dalton (1999) found that individuals given a harmful bias reported significantly more health symptoms upon exposure to an odorant than those receiving the same odorant, but no harmful effects indicated. Thus, the author concluded that prejudiced odor perceptions and reactions underscore the incredible ambiguity of odor sensation and suggest that similar nonsensory factors play a large role in people's everyday reactions to ambient odors. When odors are persistent, they can result in potential health effects.

Furthermore, odor perception has been shown to affect mood, tension, stress, depression, anger, and fatigue. These conditions could potentially lead to physiological and biochemical changes with subsequent health effects (Bolla-Wilson et al., 1988; Shusterman et al., 1988, 1991). Today the medical profession clearly indicates that stress can have serious health complications.

Although periodic odor incidences are considered a nuisance, when odors persist, there have been complaints of health effects. Nuisance lawsuits continue to increase in the United States (McGinley and McGinley, 1999a). These authors presented a conceptual model for what makes an odor episode leading to a citizen's complaint. Four odor parameters make up the hierarchy of this model: (1) character, (2) intensity, (3) duration, and (4) frequency. Character denotes the type of odor (chemical makeup) or its offensiveness. Intensity refers to the overall strength of the odor or its pungency. A mild perfume can be pleasant, but an intense one can be very annoying. Duration refers to the length of time of the odor incident, i.e., how long a receptor is subjected to the odor. The longer the duration, the more likely it is to be offensive. A person passing by a bread manufacturing plant may perceive the odor to be very pleasant. However, one living near the bakery may consider the odor to be a nuisance and annoying. Frequency refers to how often the receptor is subject to the odor incident. The more frequently the odors occur, the more annoying each experience becomes (McGinley and McGinley, 1999a).

Shusterman (2001) indicates that physical symptoms may be reported in workplaces and community settings in which odorous chemicals are present. Health complaints can include headaches, nausea, stress, eye irritation, throat irritation, cough, shortness of breath, running nose, and discomfort. Often these symptoms are the result of specific chemical compounds emitted. Although the chemical causing the odor is below known toxicological effects, if persistent, it may trigger health effects. Schusterman (1992) indicates that odors can have apparent health impacts by nontoxic logical mechanisms. At a workshop at Duke University on April 15–17, 1998, cosponsored by the USEPA and the National Institute on Deafness and Other Communication Diseases, participants identified three paradigms by which ambient odors may produce health symptoms as a result of odors from manure and biosolids (Schiffman et al., 2000; Schiffman and Williams, 2005). In the first paradigm, the symptoms are induced by exposure to odors that result in toxicological effects. The odorant is a marker for an irritant or other toxicological material. In this case, the concentration of the irritant is higher than the concentration at which the odor is first detected (Schiffman et al., 2000; USEPA, 1996). Ruth (1986) collated odor threshold data for approximately 450 chemicals along with reported thresholds of irritation. In the second paradigm, health symptoms can occur at an odorant concentration that is above odor thresholds but is not irritating. In the third paradigm, the odorant is part of a mixture that contains a copollutant that is fundamentally responsible for the reported health symptom. This is often the case with composting operations. Odors often connote to the public that the air may contain harmful bioaerosols.

Considerable evidence exists showing that irritation, rather than the odor, causes health symptoms (USEPA, 1996; Schiffman, 1998; Cometto-Muñiz and Cain, 1992; Schiffman et al., 2000). The odor serves as an exposure marker. Odorous, volatile compounds can produce an irritation, burning, headaches, and respiratory effects. This could be the result of a single compound or additive effects of several compounds (USEPA, 1996; Cometto-Muñiz et al., 1997). As the odor becomes progressively stronger, the vapors from volatile compounds will begin to cause irritation in the nose or eyes (Cometto-Muñiz et al., 1998). Individuals predisposed to respiratory ailments may be more likely to complain about health effects of odors (Horesh, 1966). Shusterman (2001) reported on a case where a pesticide, Metam sodium (sodium *n*-methyldithiocarbamate), was spilled into a waterway and hydrolyzed into H_2S and methyl isothiocyanate. Following the spill, public health officials were informed of odors by residents of communities living in towns downstream of the spill. Over 240 spill-related incidences were reported in local hospitals. Persistent health effects were reported by 197 individuals.

Haahtela et al. (1992) reported on community health problems associated with the release of hydrogen sulfide (H_2S) from a pulp mill. Deane and Sanders (1977) investigated health effects of exposure to odors emanating from pulp mills. In addition to annoyance, they reported that headaches showed a positive relationship to the amount of odor by which the respondent was bothered. There were no significant trends or relationship to doctors or hospital visits or the respondent's estimate of general health. Shim and Williams (1986) reported that a survey of sixty asthmatic patients revealed a history of respiratory symptoms in fifty-seven of the patients on exposure to one or more common odors. They concluded that odors are an important cause of worsening asthma.

Perceived health effects or "environmental worries" were observed to occur by residents near hazardous waste sites (Shusterman et al., 1991). The authors found that headaches, nausea, eye, and throat irritation reported by adult respondents had a positive significant relationship to frequency of odor perception and degree of worry.

Shusterman et al. (1991) conducted epidemiological studies evaluating the relationship between environmental odors and health symptoms near hazardous waste sites. Significant positive relationships were observed between prevalence of several symptoms (headaches, nausea, eye, and throat irritation) and both frequency of odor perception and degree of worry.

Predominantly, odors tend to be a nuisance when they are infrequent or at a very low intensity. However, when odors persist or are intense, the public often complains of public health effects. These complaints can consist of eye, nose, and throat irritations, nausea, headaches, cough, shortness of breath, and stress (Epstein, 2004). Cain and Cometto-Muñiz (2004) reviewed the literature on potential health effects of biosolids odors. They concluded the following:

- Odors do not cause signs of illness in the healthy.
- The acceptability-unacceptability of odors varies systematically and predictably with circumstances of exposure and depends upon the *meaning* people associate with the exposure.

- Below-toxic levels of exposure and symptoms associated with odors involve no pathology.
- Removal of the source of odor results in virtually immediate reduction of symptoms.
- Nonphysical variables, such as anxiety and stress, seem to mediate symptoms from odors. The authors also stated that for people with certain chronic illnesses, exposure to malodors might exacerbate existing symptoms and possibly existing signs.

Studies at composting facilities have found that the compounds measured are present in concentrations below the threshold limit values set by the Environmental Protection Agency, the Occupational Safety and Health Administration, and other worker and public health agencies. Since many of the compounds released by composting have extremely low odor thresholds, i.e., they are detectable at very low concentrations, maintenance of low odor concentrations should prevent health impacts.

There are no scientific or medical citations of odor health impacts from composting facilities.

CONCLUSION

A crucial part of planning for treatment of odors is identifying and characterizing odors. There are several strategies. These included:

- Odor survey
  - Identify sources
  - Determine emission rates
  - Determine odor detectability and odor intensity
  - Identify odor-causing compounds
  - Determine an acceptable odor standard
  - Consider any new odor generated during the treatment
- Assessment of meteorological conditions and atmospheric dispersion modeling
  - Wind speed and direction
  - Temperature
  - Inversion conditions

These should include conditions during a twenty-four-hour period, as well as during the season.

The atmospheric data and the odor survey information can then be used for the following:

- Model the impacts of odor from each source
- Evaluate potential odor mitigation efforts
- Determine the degree of odor control required to meet community odor standards

Over the years more data on odor characterization and VOC emissions have been gathered and published. This has given facility planners and operators knowledge on how to minimize odors and the inconvenience to neighbors.

REFERENCES

ASTM. 1981. *Guidelines for the selection and training of sensory panel members*. Special Tech. Publ. 758. Philadelphia, PA: American Society for Testing and Materials.

ASTM. 1999. *Standard practice for suprathreshold odor intensity measurement*. ASTM E544-99. Philadelphia, PA: American Society for Testing Materials.

AWMA. 2002. *Guidelines for odor sampling and measurements by dynamic dilution olfactometry*. Pittsburgh, PA: Air and Waste Management Association, EE-6 Odor Subcommittee on Standardization of Odor Sampling and Measurement.

Bolla-Wilson, K., Wilson, R. J., and Bleecker, M. L. 1988. Conditioning of physical symptoms after neurotoxic exposure. *J. Occup. Med.* 30:684–86.

Brenk, C., and Garibay, M. 1995. *Emission rate characterization of open windrow sludge composting operations*. Rep. no. final report. Diamond Bar, CA: South Coast Air Management District.

Cain, W. S. 1987. Indoor air as a source of annoyance. In *Environmental annoyance: Characterization, measurement and control*, ed. H. S. Koelega, 189–200. Amsterdam: Elsevier.

Cain, W. S. 2006. Perception and reality of odor. In *U.S. Composting Council's 14th Annual Conference and Trade Show*, Albuquerque, NM.

Cain, W. S., and Cometto-Muñiz, J. E. 2004. *Health effects of biosolids odors: A literature review*. Alexandria, VA: Water Environment Research Foundation.

Card, T. R., and Schmidt, C. E. 2008. *Organic material composting and drying focusing on greenwaste compost air emissions data review*. Fresno, CA: San Joaquin Valley Air Pollution Control District.

CIWMB. 2007. *Emissions testing of volatile organic compounds from greenwaste composting at the Modesto Compost Facility in the San Joaquin Valley*. Sacramento: California Integrated Waste Management Board.

Cometto-Muñiz, J. E., and Cain, W. S. 1992. Sensory irritation relation to indoor air pollution. *Ann. NY Acad. Sci.* 641:137–51.

Cometto-Muñiz, J. E., Cain, W. S., Abraham, M. H., and Kumarsingh, R. 1998. Trigeminal and olfactory chemosensory impact of selected terpenes. *Pharmacol. Biochem. Behav.* 60(3):765–70.

Cometto-Muñiz, J. E., Cain, W. S., and Hundell, H. K. 1997. Agonistic sensory effects of airborne chemicals in mixtures; odor, nasal pungency, and eye irritation. *Precept Psychophys.* 59:665–74.

Dalton, P. 1999. Cognitive influences on health symptoms from acute chemical exposure. *Health Physiol.* 18(6):579–90.

Deane, M., and Sanders, G. 1977. Health effects of exposure to community odors from pulp mills, Eureka, 1971. *Environ. Res.* 14:164–81.

Diosey, P. G. 2008. Modeling odors in the age of AERMOD. Presented at Odors and Air Emissions Conference 2008. Phoenix, AZ.

E&A Environmental Consultants, Inc. 1993. *Food waste collection and composting demonstration project*. Seattle, WA: City of Seattle Solid Waste Utility.

Elwell, D. L., Borger, D. C., Blaho, D. V., Fahrni, J. K., Keener, H. M., and Willett, L. B. 2004. Changes in concentrations of malodorous compounds during controlled aeration composting. *Compost Sci. Util.* 12:102–7.

EN 13725. 2003. *Air quality—Determination of odour concentration by dynamic olfactometry*. London: BSI.

Epstein, E. 1997. *The science of composting*. Boca Raton, FL: CRC Press.

Epstein, E. 2004. Public health issues—Odors and potentially toxic air emissions. In *Control of odors and emissions from wastewater treatment plants*, ed. T. Mahin. Alexandria, VA: Water Environment Federation.

Haahtela, T., Martilla, O., Vikka, V., Jappinen, P., and Jaakkola, J. J. K. 1992. The South Karelia air pollution study: Acute health effects of malodorous sulfur air pollutants released by a pulp mill. *Am. J. Public Health* 82:603–5.

Haug, R. T. 1993. *The practical handbook of compost engineering*. Boca Raton, FL: Lewis Publishers.

Heida, H., Bartman, F., and van der Zee, S. C. 1995. Occupational exposure and indoor air quality monitoring in a composting facility. *Am. Ind. Hyg. Assoc. J.* 56:39–43.

Hentz, L. H., Murray, C. M., Thompson, J. L., Gasner, L. L., and Dunson Jr., J. B. 1992. Odor control research at the Montgomery County regional composting facility. *Water Environ. Res.* 64:13–18.

Hino, M. 1968. Maximum ground-level concentrations and sampling time. *Atmos. Environ* 2:149–56.

Horesh, A. J. 1966. The role of odors and vapors in allergic disease. *J. Asthma Res.* 4:125–36.

Horowitz, R. 2010. Are compost emissions a strong ozone source? *BioCycle* 51:18–22.

Lees, P. S. J., and Tockman, M. S. 1987. *Evaluation of possible health impacts of WSSC Site II sewage sludge composting operations*. Report to Maryland Department of Health and Mental Hygiene. Baltimore: Johns Hopkins University, School of Hygiene and Public Health.

Kim, J. Y., Park, J. K., Emmons, B., and Armstrong, D. E. 1995. Survey of volatile organic compounds at a municipal solid waste co-composting facility. *Water Environ. Res.* 67:1044–51.

Knasko, S. C. 1993. Performance, mood, health during exposure to intermittent odors. *Arch. Environ. Health* 48(5):305–8.

Mahin, T. D. 1998. Using D/T levels. *Water Environ. Technol.* 10(2):49–53.

McGinley, M. A., and McGinley, C. M. 1999a. The 'gray line' between odor nuisance and health effects. Paper presented at Air and Waste Management Association, 92nd Annual Meeting and Exhibition, St. Louis, MO.

McGinley, M. A., and McGinley, C. M. 1999b. Odor evaluation fundamentals and applications for indoor air quality research. Paper presented at The First NSF International Conference on Indoor Air Health, Denver, CO.

Miller, F. C. 1993. Minimizing odor generation. In *Science and engineering of composting and design, environmental, microbiological, and utilization aspects*, ed. H. A. J. Hoitink and H. M. Keener. Worthington, OH: Renaissance Publ.

Peterson, M. K., Johnson, D. L., and Pleus, C. 2000. Characterization of emissions from two yard-waste composting facilities. In *Odors and VOC Emissions 2000*. Cincinnati, OH: Water Environment Federation.

Rosenfeld, P., Grey, M., and Sellew, P. 2004. Measurement of biosolids compost odor emissions from a windrow, static pile, and biofilter. *Water Environ. Res.* 76:310–15.

Ruth, J. H. 1986. Odor thresholds and irritation levels of several chemical substances: A review. *Am. Indust. Hygiene Assoc. J.* 47(A):142–51.

Schiffman, S. S. 1998. Livestock odors: Implications for human health and well-being. *J. Animal Sci.* 76(5):1343–55.

Schiffman, S. S., Walker, J. M., Dalton, P., Lorig, T. S., Raymer, J. H., Shusterman, D., and Williams, C. M. 2000. Potential health effects of odor from animal operations, wastewater treatment, and recycling of byproducts. *J. Agromed.* 7(1):7–81.

Schiffman, S. S., and Williams, C. M. 2005. Science of odor as a potential health issue. *J. Environ. Qual.* 34:129–38.

Shim, C., and Williams Jr., M. H. 1986. Effect of odors in asthma. *Am. J. Med.* 80:18–22.

Shusterman, D. 1992. Critical review: The health significance of environmental odor pollution. *Arch. Environ. Health* 47(1):88–91.

Shusterman, D. 2001. Odor-associated health complaints: Competing explanatory models. *Chem. Senses* 26:339–43.

Shusterman, D., Balmes, J., and Cone, J. 1988. Behavioral sensitization to irritants/odorants after acute overexposure. *J. Occup. Med.* 30:556–67.

Shusterman, D., Lipscomb, J., Neutra, R., and Satin, K. 1991. Symptom prevalence and odor-worry interaction near hazardous waste sites. *Environ. Health Perspect.* 94:25–30.

Van Durme, G. P., McNamara, B. F., and McGinley, C. 1990. Characterization and treatment of composting emissions at Hampton Roads Sanitation District. Paper presented at 63rd Annual Water Pollution Control Federation Conference, Washington, DC.

USEPA. 1989. *Air Superfund national technical guidance study series: Application of air pathway analysis for Superfund activities*. Vol. 1, interim final, EPA-450/1-89-001. Washington, DC: U.S. Environmental Protection Agency, Office of Air Quality Planning and Standards.

USEPA. 1996. *Air quality criteria for particulate matter*. Vols. I–III, Rep. EPA/600/P-95/001a,b,c,F. Research Triangle Park, NC: National Center for Environmental Assessment, U.S. Environmental Protection Agency, Office of Research and Development.

Walker, J. M. 1993. Control of composting odors. In *Science and engineering of composting: Design, environmental, microbiological, and utilization aspects*, ed. H. A. J. Hoitink and H. M. Keener. Worthington, OH: Renaissance Publications.

WEF. 2004. *Control of odors and emissions from wastewater treatment plants*. WEF manual of practice 25. Alexandria, VA: Water Environment Federation.

Wilber, C., and Murray, C. 1990. Odor source evaluation. *BioCycle* 31:68–72.

Williams, T. O., and Miller, F. C. 1993. Composting facility odor control using biofilters. In *Science and engineering of composting: Design, environmental, microbial, and utilization aspects*, ed. H. A. J. Hoitink and H. M. Keener. Wooster, OH: Ohio State University.

10 Operational Control of Odors

INTRODUCTION

Operational control is a key to good odor management. It is relatively easy, although costly, to control odors from an enclosed facility. Principally, this can be accomplished using biofilters or other odor control systems, such as chemical control. It is more difficult to control odors from open systems, such as windrow or aerated static pile (ASP). For these systems, operational control is paramount.

The first step in operational odor management is in the planning process and design of a facility. This recommendation was pointed out in earlier chapters and involves site selection, system selection, and the design of the facility.

The basic principles in operational control management involve the following:

- Attention to climatic and microclimatic conditions
- Management of the carbon-to-nitrogen ratio
- Aeration and oxygen
- Control of moisture
- Control of temperature
- Maintaining proper physical structure of the matrix (mixing and porosity)

A valuable tool to determine corrective actions with respect to odors is an odor survey. This requires identifying the potential odor sources and trying to minimize or eliminate them. It also can involve identifying receptors that may be impacted and when.

To identify specific local potential odor sources, other than overall composting odors, requires a survey. The survey can consist of identifying site odors, feedstock delivery and deposit odors, storage and delivery odors, and composting and curing odors. The survey should identify the types of odors (see Chapter 9), and odor intensity as related to the activity.

An example of a comprehensive odor survey for a biosolids windrow operation is shown in Table 10.1. Table 10.2 shows a comprehensive survey for ASP after being changed from a windrow operation (Epstein and Wu, 2000).

For a windrow operation the major sources of overall odors were:

- Curing—61.6%
- Composting—27.1%
- Postprocessing—7.7%

TABLE 10.1
Sources and Magnitude of Odors during Windrow Operations

Source	Odor Concentration (D/T)	Duration h/day	Total Emissions (10^3 odor units/day)[a]	Percentage
Feedstock				1.0
Delivery	200	8	77	
Storage	386	24	14,895	
Transfer	82	8	26	
Feedstock mixing				
Mixing	1,500	8	468	1.0
Mix storage	386	24	14,895	
Mix pile transfer	82	8	26	
Composting				27
Pile construction	82	8	26	
Surface (1–5 days)	1,370	24	169,174	
Turning (1–5 days)	5,460	0.8	10,319	
Surface (6–10 days)	1,500	24	185,227	
Turning (6–10 days)	7,080	0.8	13,381	
Surface (11–20 days)	23	24	2,840	
Turning (11–20 days)	5,000	0.8	12,600	
Surface (21–28 days)	89	24	10,990	
Turning (21–28 days)	3,000	0.8	7,560	
Pile teardown	82	8	26	
Curing				61.6
Surface (1–7 days)	7,080	24	455,350	
Surface (8–28 days)	7,080	24	455,350	
Surface (29–70 days)	177	24	22,767	
Pile teardown	7,080	8	2,209	
Postprocessing				7.7
Screening	1,000	8	312	
Storage	1,000	24	115,767	
Transfer	1,000	8	312	
Standing water				1.7
Compost runoff	149	24	3,833	
Curing runoff	149	24	17,249	
Agitated curing runoff	12,800	24	3,994	
Total			1,519,672	

Note: Total emissions are a function of odor concentration, duration, surface area or plume size, and the flux rate.

TABLE 10.2
Sources and Magnitude of Odors during ASP Operations after Being Changed from a Windrow Operation

Source	Odor Concentration (D/T)	Duration (h/day)	Total Emissions (10^3 odor (units/ day	Percentage
Feedstock delivery	200	8	77	<0.0
Feedstock mixing	1,500	8	468	<0.0
Composting				94.9
Pile construction	82	8	26	
Surface (1–5 days)	648	24	403,685	
Surface (6–14 days)	449	24	111,886	
Surface (15–28 days)	82	24	14,303	
Pile teardown	82	8	26	
Curing				3.3
Surface (1–7 days)	82	24	7,802	
Surface (8–28 days)	82	24	2,601	
Surface (29–70 days)	82	24	7,802	
Pile teardown	82	8	26	
Postprocessing				1.7
Screening	82	8	26	
Storage	82	24	9,493	
Transfer	82	8	26	
Total			**558,247**	

- Standing water (site conditions)—1.7%
- Feedstock delivery and storage—1%
- Feedstock mixing—1%

There are several aspects that can be pointed out from these data. Different locations or sources produce different types of odor. Although feedstock odors were relatively low, odors from untreated biosolids, grass, or decayed foodwaste can be much more offensive than curing odors. Curing odors, although strong because of the large surface area, may not be very offensive, as the compost or earthy odor may be more acceptable to potential receptors. Furthermore, Tables 10.1 and 10.2 show that site odors can be important and should be easily eliminated. The postprocessing odors were high due to the length of storage time. Surface odors from curing were very high during twenty-four hours for the first twenty-eight days. In Table 10.1, 27% of the odors were generated during the windrow composting period. Compost was not stabilized before being placed in curing, and the curing generated 62% of on-site odors. Although turning gave rise to emissions with much higher odor concentrations, turning was only conducted for short periods of time. Eighty-nine percent of the odors generated from composting and curing piles were from surface emissions from quiescent windrows.

Similar data were reported by Iacoboni et al. (1980) at the Los Angeles County Sanitation District. Iacoboni et al. (1980) showed that 85% of composting odors are emitted from windrows in their ambient or unturned state. Brenk and Garibay (1995) reported that emissions of ammonia and methane were greatest during the first twenty-two days of the composting cycle.

In the case of the biosolids ASP operation, odor generation was considerably different. In addition to changing the composting system, materials handling and site conditions were improved, resulting in an overal reduction of odors fom 1,519,672 total emissions to 558,247 (10^3 odor units per day). Ninty-five percent of the odors were generated from the composting piles, and only 3% were generated from aerated curing. This is due to the fact that compost stabilized much more quickly under an aerated system, as indicated by lower temperature. Therefore, there was less potential for odor generation from the curing piles. The overall generation of odor was one-third the odor generation from the windrow system. In addition, odor emissions from an ASP could be captured and treated using a biofilter or by using positive air and e-PTFE micropore compost cover.

Several studies have shown that odor emissions are greatest during the first seven to fourteen days of composting. Therefore, it is recommended that during that period the ASP sytem should be operated in a negative mode with the air captured in a biofilter or other odor control system. This would also be the period where a fabric cover would be most effective.

In the odor sources, as shown in Tables 10.1 and 10.2, intensity and its offensiveness are not considered. For example, while new compost piles may not produce the highest number of odor units, the intensity of the odor generated may be higher because of the type of compounds formed during the early stages of composting. Higher-intensity odors are detectable at lower concentration, and therefore have a relatively higher potential to cause odor impacts.

A study on emissions of volatile organic compounds (VOCs) from greenwaste composting showed that the majority of the VOC emissions occurred during early stages of composting. Approximately 80% of the VOCs from greenwaste and 70% of the VOCs from food waste were emitted during the first two weeks of composting. A surface emission survey suggested that close to 85% of the emissions occurred from windrow tops as compared to the sides of the windrows (CIWMB, 2008).

In a study by Murray and Thompson (1986) at Site II in Maryland, which used the aerated static pile, the area sources of odors were as follows:

Area Sources	Percentage Contribution from Each Pile (%)
Compost	51.7
Drying pile	16.2
Curing pile	11.1
Compost storage	10.8
New chips storage	3.5
Used chips storage	2.9
Mixing area	2.0
Drying pile under construction	0.5
Screened compost	0.1
Screened chips	1.1

Knowledge obtained from surveys can be used to reduce odors. Improvements such as housekeeping, mixing, screening, and storage areas should properly be managed and their odor sources eliminated. Granted that these measures do not address the larger sources of odors from composting and curing, they still have an impact on overall odor emissions. One must understand that reaction to odors can often be triggered by one small source. A good example, such as curing, due to the large amount of material on-site, provided the largest source of odors. This odor may not have been an offensive odor, and nearby receptors did not complain. However, bringing in an offensive feedstock into an uncontrolled mixing area could trigger complaints and result in a citation or regulatory action. Therefore, it is imortant to take care of those areas that can produce odors immediately and can be easily managed or controlled. Concurrently, effort on minimizing or eliminating odors from the more difficult sources must be accomplished by operators.

Comprehensive odor literature sources are:

Buyuksonmez, F., *Comprehensive Compost Odor Response Project*, Publication 442-07-001, San Diego University, Integrated Waste Management Board, Sacramento, CA, 2007

Odor Control in Wastewater Treatment Plants, WEF Manual of Practice 22, ASCE Manuals and Reports on Engineering Practice 82, Alexandria, VA, 1995

PRINCIPLES OF OPERATIONAL CONTROL

Attention to Climatic and Microclimatic Conditions

The climate and microclimate principally affect the dispersion of odor and VOC emissions. This aspect was discussed in Chapter 9. An excellent reference that discusses meteorology as related to composting is Haug (1993). In addition to odor dispersion and potential impacts to the surrounding area is the dispersion of dust and bioaerosols.

Once a facility is established, little can be done to modify the climate. Some modification of the microclimate can be achieved using the concept of shelterbelts. Shelterbelts consist of rows of trees that break up the flow of air. They create an obstacle for moving air masses. Shelterbelts force turbulent fresh air up and over the tree row, and they will also moderate and evenly distribute a more gentle airflow through the trees. It is believed that shelterbelts have the ability to lift some of the odor plume into the lower atmosphere, where winds aloft mix and dilute the odor. The greatest dilution of odor occurs above and downwind from the quiet zone created by the action of wind passing over the shelterbelt. Beyond the quiet zone, more fresh air and less odorous air return to the ground, thereby reducing the movement of compost odors off-site and resulting in a more dilute air.

Shelterbelts (Figure 6.1) create a physical barrier to wind and air movement. The trees absorb wind energy, which reduces its speed near the ground. As a result, the air near composting facilities will pick up fewer dust particles and less odorous gases.

Shelterbelts can also provide a visible barrier. It has often been stated that odors can be seen. When receptors see a plume of steam coming from a windrow operation during windrow turnings, they often assume that toxic emissions are being released.

They do not realize that the plume is primarily steam, i.e., water vapors. A shelterbelt or other visual barriers can eliminate this problem.

Relatively little can be done to modify the climate or microclimate. Management of the facility must give attention to operating the facility in a manner that would reduce any potential impact to nearby residences. One of the less expensive devices that could provide very valuable meteorological data to the operator is a weather station that provides information on wind direction, speed, and other climatic data. This information can provide data as to the best windrow turning times. It can also be valuable to negate wrong odor complaints. In one case, it was demonstrated to regulators and legal personnel that the day the community complained of odors, the wind was blowing in the opposite direction.

Attention also needs to be given to weather inversions. Temperature inversions refer to an increase of temperature with height; they result in very stable air conditions. The normal vertical temperature gradient is inverted. Thus, the air near the surface is cooler. Under this condition, pollutants, including odors, can be trapped and released as the earth warms up. Windrow turning during that period can result in excessive odors until the inversion is released.

The impact of climatological information, especially wind conditions on a community of receptors, was described in a report by TRC regarding the Cedar Grove Composting Facility in Seattle, Washington (TRC, 1991).

This emphasizes the need to manage composting operations, especially turning and pile breakdown in relation to climatological conditions.

Management of the Carbon-to-Nitrogen Ratio (C:N)

The two most important microbial nutrients are carbon and nitrogen. Microorganisms use carbon for energy and growth and nitrogen for cell formation and protein synthesis. Not all the carbon in organic matter is available to the microorganisms. Thus, a measurement of total carbon is not a good indicator of the available carbon (Epstein, 1997; Das, 2000). There have been numerous attempts to measure the available carbon. There is no single biological, physical, or chemical fractionation technique developed that adequately describes the continuum of organic carbon that exists in nature. In the past, the measurement of organic carbon in soil, which much closely resembles available carbon, was done by rapid dichromate oxidation. The preferred method was the Walkley-Black procedure (Nelson and Sommers, 1982). This still may be a good approximation for the determination of available carbon for biological decomposition of organic matter. Others prefer total carbon determination by incubation. Lignin is one component of organic matter that is not readily biodegradable. Since it is carbonaceous, when measuring total carbon it gives an erroneous indication of the available carbon.

Tchobanoglous et al. (1993) suggested the determination of available carbon to be based on lignin and provided the following equation:

$$\text{Biodegradable fraction (BF)} = 0.83 - (0.028) \times \text{lignin content of the volatile solids as a percent of dry weight}$$

The numbers 83 and 0.028 are empirical constants.

Paula et al. (2006) reviewed the literature and analyzed the data to determine the feasibility of using a combination of acid hydrolysis and CO_2–C release during long-term incubation to determine soil organic carbon (SOC) pool sizes.

During microbial growth, approximately twenty-five to thirty parts of carbon are needed for every unit of nitrogen (Epstein, 1997). When carbon is excessive or nitrogen is low, the available carbon is not fully utilized. Thus, the excess nitrogen is lost to the atmosphere as ammonia, amines, or nitrous oxide, which can result in odors. Several facilities have been closed when excessive grass or biosolids, which have a low C/N ratio, are being composted without having sufficient carbon.

Aeration and Oxygen

Oxygen must be available for the composting process to proceed. Large quantities of oxygen are utilized during the composting process. Oxygen is provided through aeration. At low oxygen levels, anaerobic conditions can occur with the release of odorous compounds such as methane, organic acids, hydrogen sulfide, and other compounds. Generally, oxygen levels below 5% will result in anaerobic conditions. It is best to maintain oxygen levels at 10% or greater. Because of the high demand for oxygen by the microbial activity, oxygen depletion can be very rapid, especially early in the composting process.

Aeration is provided in several different ways. In windrow composting, most of the aeration is the result of convective air, except during turning. As the windrow pile heats up, air is drawn into the center of the pile, which is hot, and leaves through the top of the pile. This is similar to the occurrence in a chimney. Under passive aeration, perforated pipes in the pile draw outside air into the warm center portion of the pile. With ASP and other technologies, blowers are used to provide aeration. Empirically, it has been determined that in order to control temperature and moisture and to supply sufficient oxygen, a value of 76 to 129 m^3h/dry tonne (3,000 to 5,000 cfh/dry ton) is needed for biosolids with a wood chip bulking agent. Low-level aeration can be supplied continuously, or aeration can be supplied intermittently. In any event, it is important that the oxygen levels do not drop below 10%.

Moisture

Moisture is essential for microbial activity. It can affect the rate of decomposition. At moisture contents below 40%, microbial activity is reduced. At moisture contents greater than 60%, the effective pore space or free pore space is reduced since the pores are filled with water (Epstein, 1997; Rynk et al., 1992). During composting, the moisture content changes because of water evaporation. A key aspect in composting is to proceed as rapidly as possible toward a stabilized product. Emissions from a stabilized product are usually not offensive, as the odor of stabilized compost is that of a rich soil.

Temperature

The progression of temperature during the composting process is from mesophilic (10 to 41°C, 50 to 105°F) to thermophilic (> 41°C, >105°F) and back to mesophilic. Temperature control is important to destroy pathogens at 55°C (131°F) and weed seeds. The critical temperature for destroying most weed seeds is 63°C (145°F). Therefore, it is best to control temperatures to no more than 65°C (149°F). At high temperatures, VOC emissions are higher. Decomposition of the organic matter and reaching stabilization are faster at mesophilic temperature. Therefore, it is important to achieve thermophilic temperature early in the process to control pathogens according to the regulatory requirements and then proceed to achieve mesophilic temperatures in the 45 to 50°C (113 to 122°F) range.

Maintaining Proper Physical Structure of the Matrix

The physical structure, primarily porosity, is an important integral part of the composting matrix. Porosity is a measure of the air space. As indicated, the free pore space, which allows oxygen to get to the organisms, is important for maintaining aerobic conditions. Maintaining the proper porosity is achieved through particle sizing and the use of amendments or bulking agents. Feedstocks such as biosolids and food waste are very dense and do not have the proper structure. They also have high moisture content, and therefore will require a bulking agent such as wood chips to provide the proper porosity and moisture content.

MANAGERIAL ASPECTS OF ODOR MANAGEMENT

Odor mitigation for a windrow operation is much more difficult to achieve than for ASP or enclosed facilities, where emissions can be captured and treated or controlled, principally by using a biofilter or the use of covers, as discussed in Chapter 11.

The odors from feedstock delivery and storage, feedstock mixing, and postprocessing, which often occur in an enclosure, may be controlled using a biofilter. Even if the area where these activities take place is not enclosed, odor control can be achieved. Just having the location of feedstock deposit under negative air sent to a simple compost biofilter is very helpful. Site odors may be eliminated by avoiding runoff and ponding. The use of lime can reduce odors from standing water. Spillage of feedstock material should be picked up.

With respect to composting and curing odors, several measures can be attempted to reduce odors. Turing of windrows should be avoided when the wind is in the direction of receptors or a majority of the receptors. As pointed out, one of the least expensive and very valuable pieces of equipment on a site is a weather station. This can provide the supervisor or operators the optimal time to turn windrows. In locations where inversions occur, windrow turning should be postponed until atmospheric dispersion conditions improve.

During turning, using either a very fine water spray or odor counteractants reduces emissions. Several facilities use a sprayer mounted on a truck that travels along the windrow. Many turner manufacturerers provide water tanks and sprayers on the turner. One facility that was observed had a trailing hose on a reel mounted on the turner providing spray (Figure 10.1). Covers can be used over windrows or static piles (Figure 10.2). To reduce odors during curing or from storage piles that are not turned or moved, the use of covers as shown in Figure 10.3 may be realistic. The use of covers is discussed in greater detail in Chapter 11.

Controlling odors in ASP systems is easier than with windrows. When an ASP is operated under negative air, the exhaust can be sent to a biofilter or to another odor control system. Another possibility is the use of covers, as shown in Figure 10.2. If the piles are in positive air, covers can also be used. The use

FIGURE 10.1 Spraying water over a windrow during turning. Notice the darker color of the windrow that received the spray.

FIGURE 10.2 Using a cover for odor control. (Courtesy of Managed Organic Recycling, Inc.)

FIGURE 10.3 Use of cover during curing to reduce surface emissions and reduce evaporation in Santa Rosa, California.

of a mini-biofilter can provide some odor control. Epstein et al. (1976) applied 30 cm (12 inches) of screened compost over negative aerated static piles, both to retain temperature within the piles and to reduce surface emissions. Further studies showed that a blanket of 45 cm (18 inches) of unscreened compost was also effective. A study by the CIWMB (2008) found that the application of a finished compost blanket resulted in 82% reduction in VOC emissions during the first seven days and 75% reduction for the first fourteen days of composting, compared to the control.

Previously it was indicated that an odor survey can provide very usefull information for better odor management. The following is an example of an odor evaluation and resultant program based on the findings.

The town of Islip, New York, had an odor evaluation program regarding its yard waste composting site (ADL, 1989). It found the following significant items:

- Significant odors thought to originate from windrow turning and grass shredding were found within 1.2 km (0.75 mile) of the site.
- Residents living in areas 0.5 to 1 km (3 to 0.6 miles) observed significant odors during windrow turning operations.
- The odor released from the pore spaces (core) of the windrows was the principal off-site odor and not surface odors.
- The potential odor release from the core of the windrows during turning was lower by a factor of 40% from day 6 to day 66 of composting.
- The greatest potentials for off-site odor nuisance as related to atmospheric dispersion were:
    - Windspeed 8.0 km/h (5 mph)
    - Direction SSW
    - Stability class D
    - Receptor distance 0.5 km (0.3 mile)

- The Sentometer® underestimated the strength of odors in the vapor plume during windrow turning by up to a factor of 10. Measuring reduced sulfur using a Jerome analyzer proved best suited for chemical measurements to predict compost odor strength.

The following were some of the recommendations:

- Restrict windrow turning of the most odorous windrows to one hour after sunrise to one hour prior to sunset, when atmospheric dispersion is maximized.
- Consider changing the process from mechanically turned windrow to static pile with forced down-draft (negative) aeration.
- Testing of bulking ratio should be done to determine what ratio may be best to reduce odors.
- Site improvements:
  - Consider rearranging the windrows from the current orientation to avoid wind-induced stripping of windrow odors.
  - Odors from unused areas, recharge basins, and grass areas, which are potential odor sources, should be treated with lime, hypochlorite, or other oxidents.

- Process control:
  - Maintain a C:N ratio of 40:1 by the addition of a carbon source.
  - Moisture control—avoid anaerobic conditions.
  - Piles should be prevented from overheating—too high temperature can cause the release of odorous compounds. Avoid temperatures exceeding 60°C.
  - Develop an aeration scheme (e.g., turing frequency) to maintain aerobic conditions.

Islip incorporated many of these changes, which greatly reduced odors (Buckner, 1996, 2002). These studies showed that low odor levels were generally maintained only in windrows with mean oxygen concentrations of approximately 10% or more. Oxygen concentration was a key factor and was correlated with odor levels. Buckner (2002) concluded from the studies that an essential component of the solution to odor control is process management.

It is obvious that the conditions and modifications indicated in the Islip case do not apply to all facilities in all locations. However, the principles apply.

Process optimization and improved operating procedures may provide significant reduction in odor emissions. It is evident from both the windrow and static pile data that process technology and operations will affect odor production in different ways. Conventional windrow systems appear to produce more odor emissions during the curing phase, whereas in the ASP system emissions are low during curing due to a more stable product. Some of the operational changes and capital improvements, such as providing a means of forced aeration for an ASP system, could result in substantial additional costs. The following shows how improvements can be made to reduce odors:

- If mixing is done with a front-end loader, consider the purchase of a batch mixer or the use of continuous conveyor feed-mixing equipment for larger facilities.
- If compost piles are too hot (>165°F, 74°C), increase aeration; try to keep temperatures below 145°F (63°C).
- If methane is detected, increase aeration so that oxygen levels are over 10%.
- Reduce surface emissions of curing pile by using slow, positive aeration, or negative aeration to a biofilter.
- Evaluate moisture content. If too wet, increase aeration; if too dry, add water.
- Evaluate bulking agent ratio. Increase amount of carbon for highly decomposable feedstocks.
- Measure respiration rate, temperature, and oxygen of curing piles. If piles are too hot or have low oxygen or high respiration rate, consider placing curing piles under aeration. Alternatively, increase mixing or turning.

In a Denmark facility, windrows were placed over a negative aeration system that withdrew the air and sent it to a biofilter. This significantly reduced odors.

It is obvious that the conditions and modifications indicated in the Islip, New York, case do not apply to all facilities in all locations, but the principles apply.

COVERS

Two types of covers have been used. One type is often referred to as a blanket cover. It consists of compost, wood chips, and other organic materials placed over piles. This concept was first used at Beltsville, Maryland (Epstein et al., 1976). Thirty centimeters (12 inches) of screened compost or 45 cm (18 inches) of unscreened compost was placed over aerated static piles. The material was used both as an insulation and to control surface emissions.

Recently, there was further evaluation of blanket covers (Buyuksonmez and Yucel, 2006; Buyuksonmez, 2007; CIWMB, 2008). Blanket covers were found to be an effective measure for reducing odor emissions and VOCs. Compost blankets significantly reduced terpenes, ammonia, and sulfur compounds. The results with mercaptans were inconclusive since they were not found in most samples. The blanket cover was ineffective in reducing organic acids. In the CIWMB (2008) Comprehensive Compost Odor Response Project, compost cover resulted in substanial odor reduction ranging from 51.5 to 98% on the total terpene emissions for various application rates. Ammonia emissions were greatly reduced initially but decreased with time. During the first hour the reduction in ammonia was 86% (CIWMB, 2008). This was a laboratory study with small samples and may not represent emissions occurring in the field with a very large amount of material.

In another report by the California Integrated Waste Management Board (CIWMB, 2008), the application of the finished compost blanket (pseudobiofilter) resulted in an 82% reduction in VOC emissions for the first seven days and 75% reduction for the first fourteen days. It was much more effective than the use of an additive (Table 10.3). The report indicated that the compost blanket cover was also a very significant cost reduction when using it vs. an additive. These data were for a specific additive and could vary with others.

TABLE 10.3
Comparison of VOC Emission Reductions from Greenwaste Using an Additive or Blanket Cover

Period	Additive	Blanket Cover (Pseudobiofilter)
7 days (week 1)	42%	82%
14 days (first 2 weeks)	14%	75%

Source: CIWMB, *Emissions Testing of Volatile Organic Compounds from Greenwaste Composting at the Modesto Compost Facility in the San Joaquin Valley*, California Integrated Waste Management Board, Sacramento, 2008. With permission.

A second type of cover consists of a somewhat porous fabric, as shown in Figures 10.2 and 10.3. The material is permeable to oxygen but relatively impermeable to water. It not only minimizes odors, but also reduces the emissions of volatile organic compounds and, to some extent, ammonia. Moisture condenses on the underside of the fabric and traps odorant compounds and volatile organic compounds.

Recently several studies predominantly evaluated the use of e-PTFE fabric covers (OS&E, 2005; CH2MHill, 2005; MOR, 2007, 2009). Unfortunately, these studies have not been published or reported in *BioCycle* or similar journals. There are no data in referred journals. The following are some exmples of the results.

In 2005, OS&E (2005) conducted a study to evaluate the Gore™ cover system on aerated static piles (ASPs) in comparison with an existing windrow composting for controlling odors from greenwaste and biosolids. In the OS&E study, an evaluation was made of the GORE system using a semipermeable e-PTFE membrane technology.The GORE was found to be effective in reducing odors from greenwaste. The odor removal efficiencies for greenwaste during constant aeration on the first two days of testing, based on odor concentrations below and above the cover, were 96.3% on the first day and 93.5% on the second day of testing. However, when 25% biosolids were added to the greenwaste, the odor efficiency removal dropped to 73% for the first day and 54.5% for the second day (Table 10.4). The effeciency removal rate was much lower for odorous compounds produced during the composting of biosolids and yard waste. Thus, it appears that sulfur and nitrogen compounds are able to penetrate the Gore cover.

The data in Table 10.5 showed that, even in this worst-case scenario, odor emissions from greenwaste ASPs with Gore would remain below the emissions from the presently used open greenwaste windrows. The windrows were aerated by natural draft as a result of heating in the core of the windrow (convective airflow). Odor emissions from windrows are even higher when the windrow is turned. The peak odor emissions from the mixture of greenwaste and biosolids exceeded emissions from the quiescent open windrow by a factor of close to 3.5. There was no description of the ASP. Was the exhaust treated by a biofilter or was positive air used? All

TABLE 10.4
Odor Emission Rate Using the GORE e-PTFE Membrane on Continous Aerated ASP

Day 1	Odor Concentration (D/T)[c]	Percent Reduction	Odor Emission Rate (OU/min)[d]
G100[a] inlet	6,732		
G100 cover outlet	252	96.3	65,675
G75[b] inlet	7,446		
G75 cover outlet	2,025	73	171,142
Day 2			
G100[a] inlet	3,723		
G100 cover outlet	241	93.5	14,933
G75[b] inlet	4,131		
G75 cover outlet	1,906	54	295,530

Source: OS&E, *Evaluation of Gore™ Cover System with a Semi-permeable e-PTFE Membrane Technology for Control of Odors from Greenwaste and Biosolids Composting*, Odor Science & Engineering, Inc., Bloomfield, CT, 2005. With permission from City of Oceanside, CA.

[a] 100% greenwaste.

[b] 75% greenwaste and 25% biosolids.

[c] D/T = dilution to threshold.

[d] OU/min = odor units per minute per square foot of emission surface. (= D/T × CFM)

TABLE 10.5
An Evaluation of the Gore Cover for Reducing Odors from Greenwaste and a Mixture of Greenwaste and Biosolids

Feedstock	6/3/05	6/7/05	6/9/05	Average	Adjusted for Active Surface[a]
100% greenwaste aerated; Gore ASP	38,450	14,933		26,692	26,692
100% greenwaste quiescent; Gore ASP			6,029	6,029	6,029
Greenwaste and biosolids aerated; Gore ASP	171,142	295,530		233,336	233,336
Greenwaste and biosolids quiescent; Gore ASP			17,292	17,292	17,292
Windrow (control)		203,203	197,720	200,462	66,821

Source: OS&E, *Evaluation of Gore™ Cover System with a Semi-permeable e-PTFE Membrane Technology for Control of Odors from Greenwaste and Biosolids Composting*, Odor Science & Engineering, Inc., Bloomfield, CT, 2005. With permission from City of Oceanside. CA.

[a] OU/min/ft^2 = odor units per square foot of emission surface (= (D/T) × CFM/ft^2).

the values reported are very high and could result in odor problems to the surrounding community. Table 10.6 shows the odor concentration in D/T for 100% greenwaste, and the 75% greenwaste and 25% biosolids using the Gore ASP system. The Gore system was excellent for the 100% greenwaste, but was much less effective for the 75% greenwaste and 25% biosolids. There were only 72.85% removal on day 1 and 53.9% removal on day 2. This suggests that the fabric does not remove some odorous compounds generated by biosolids. Furthermore, the D/T values were very high even with the Gore system. If these values were near the property line, it may be very offensive to residents nearby or exceed regulatory criteria.

An evaluation of the Gore at the Cedar Grove composting facility in Everett, Washington, reported a capture efficiency of 99.97% on one day. The destruction efficiency (DE) estimated how efficient the cover is at trapping and reducing emissions beneath the cover rather than allowing the emissions to be vented to the atmosphere. The DE values ranged from 80.2 to 97.0% over a four-day period (CH2MHill, 2005).

The study by Schmidt et al. (2009) indicated that GORE covers were able to meet the strict VOC emissions control goals for greenwaste as set by the San Joaquin Valley Air Pollution Control District (SJVAPCD) Draft Rule 4566, as well as for South Coast Air Quality Management District (SCAQMD) Rule 1133 for biosolids co-composting (biosolids and wood chips or greenwaste). Their studies showed better than 90% control of fugitive air emissions from compost operations for biosolids–wood chip, biosolids-greenwaste, and greenwaste mixtures. However, due to weather conditions, cold ambient temperatures, and (frozen) biosolids, the compost mix was

TABLE 10.6
Odor Concentration and Dilution Threshold (D/T) Using a Gore Cover on ASP with Greenwaste and Greenwaste with Biosolids

Time and Condition	Odor Concentration D/T	Percent Removal
Day 1		
100% greenwaste; Gore ASP inlet	6,732	
100% greenwaste; Gore ASP outlet	252	96.3
75% greenwaste, 25% biosolids; Gore ASP inlet	7,446	
75% greenwaste, 25% biosolids; Gore ASP outlet	2,025	72.8
Day 2		
100% greenwaste; Gore ASP inlet	3,723	
100% greenwaste; Gore ASP outlet	240.7	93.5
75% greenwaste, 25% biosolids; Gore ASP inlet	4,131	
75% greenwaste, 25% biosolids; Gore ASP outlet	1,905.7	53.9

Source: OS&E, *Evaluation of Gore™ Cover System with a Semi-permeable e-PTFE Membrane Technology for Control of Odors from Greenwaste and Biosolids Composting*, Odor Science & Engineering, Inc., Bloomfield, CT, 2005. With permission from City of Oceanside, CA.

Note: D/T = dilution to threshold.

slow to achieve operating temperature. The pile did not achieve the 131°F (55°C) core temperature until five days later. The pile operating temperature was not achieved until ten days after initial covering. This could significantly affect VOC emissions. VOC emissions are highest when temperatures are in the thermophilic range.

The effectiveness of covers for removal of VOCs is based on the condensation of a water layer beneath the fabric. In some conditions, if condensation does not occur or is very low, VOC emissions may not be significant.

The two rules referred to above are very strict and are summarized below:

SCAQMD 1133.1	80% VOC mitigation compared to baseline factor of 4.75 lb/ton for greenwaste and 1.78 lb/ton for biosolids composting	
SJVAPCD 4566 (Draft 2008)	Implement at least five class 1 mitigation measures in addition to one class 2 mitigation measure for active composting.	Maintain an oxygen concentration of ≥5% in every active and curing pile. Maintain the moisture content between 50 and 60%. Cover all active compost and curing piles within three hours of each turning with a waterproof covering. Maintain active compost in the temperature range of 130 to 140°F. Implement an alternative class 2 mitigation measure that demonstrates ≥80% reduction in VOC emissions.

Although it is commendable that these California districts want to reduce VOC and odor emissions from composting piles, it is clear that in the SJVAPCD regulations the authors of these rules do not understand the composting process and its reality. Furthermore, excessive regulations result in excessive costs and discourage composting.

The stated requirement to maintain temperatures—"Maintain active compost in the temperature range of 130 to 140°F"—is unrealistic. These temperatures cannot be achieved throughout the entire active period and should not be achieved. As the decomposition of readily available carbon is depleted and the compost is stabilized, temperatures will be reduced from thermophilic to mesophilic. Moreover, the thermophilic temperatures of 130 to 140°F (55 to 60°C) are less conducive to the decomposition of the organic matter and its stabilization. At those high thermophilic temperatures, more VOCs will be stripped off composting piles for a longer period. The faster stabilization and decomposition of the organic matter takes place, the fewer emissions of odors and VOCs.

The objective for having temperatures in the thermophilic range (130 to 140°F, 55 to 60°C) is to destroy pathogens. There is no timeframe for these criteria in the SJVAPCD 4566 proposed regulations. The USEPA 40CFR503 regulations designed to reduce pathogens and vector attraction in sewage sludge were based on time-temperature. These regulations and the earlier 40CFR257 were the result of studies by Dr. Burge and this author conducted at Beltsville, Maryland, in 1974–1978 and concurrent studies conducted by Sandia Laboratories.

The use of covers avoids the need to continuously replace the compost covers or pseudo-covers every time the windrow is turned.

In a study in Utah, odor reduction using MOR e-PTFE micropore compost covers reduced odors during an eighteen-day composting period from 68% to 76%. The covers were used over piles of 20% biosolids and 80% wood chips. The data are shown in Table 10.7.

In another study, ammonia emissions were greatly reduced using two different covers. Table 10.8 shows the data. Both covers showed excellent removal efficiencies ranging from 53 to 98%. The best occurred with the e-PFTE with a compost biofilter. The marine cover without a biofilter reduced ammonia emissions by 77%. VOCs were greatly reduced. The e-PTFE cover provided 84% removal and the marine cover 86% removal. Odor removal was also very efficient, with the marine cover resulting in 73% removal efficiency.

The removal efficiencies were:

- e-PFTE cover with a biofilter—98.2%
- e-PTFE cover without a biofilter—66%
- Marine cover—85.9%

These data clearly show that fabric covers are effective in reducing odors, VOCs, and ammonia. The combination of a compost blanket and e-PTFE membrane should be a very excellent odor and VOC reduction method.

TABLE 10.7
Odor Emissions Using the MOR e-PTFE Cover over a Mixture of 20% Biosolids and 89% Wood Chips

Time of Sampling	Condition	Dilution to Threshold (D/T)	Percentage Reduction (%)
Day 1	Uncovered	16,250	
	Covered	3,850	76
Day 2	Uncoveered	3,350	
	Covered	1,000	70
Day 3	Uncovered	7,750	
	Covered	2,500	68
Day 4	Uncovered	1,535	
	Covered	315	79
Day 9	Uncovered	835	
	Covered	250	70
Day 18	Uncoveered	2,570	
	Covered	805	69

Source: Courtesy of Managed Organic Recycling, Inc.

TABLE 10.8
Effect of e-PFTE Covers on Ammonia Concentration

Description	Ammonia Concentration mg/m²	μg/m²/min Ekt	lb/h/ft²	lb/ft²/h/1,000 ft² of Pile Surface Area	lb/h
e-PFTE cover with biofilter	2.5	19.231	2.3637E-07	0.0002364	0.00234
e-PFTE cover	66	507.692	6.2401E-06	0.00624	0.00618
e-PFTE no cover	140	1076.92	1.3237E-05	0.0132365	0.01310
Marine cover	20.5	161.538	1.9855E-06	0.0019854	0.00371
Marine no cover	90	984.615	1.2102E-5	0.0121020	0.02263

Source: Courtesy of MOR.

Since equipment is now available to remove and replace membranes or covers over windrows, under certain conditons their use is very appropriate and advisable.

Data showing that covers reduce odors and VOCs by greater than 90% is highly unlikely. If one covers piles with a tarp or encloses the piles in a building, a 90+% reduction can be obtained—especially with windrows, if one considers the time the covers are off. This author has done several measurements and has found better than 75% reduction in odors and VOCs. This is excellent. Maintaining covers on piles during the first seven to fourteen days greatly reduces emissions and potential odors.

Covers appear to be excellent in reducing odors and VOCs by over 75%. They are particularly effective under the following conditions:

- Windrows—machines are available for placing and removing covers
- Passive aerated piles
- Aerated static piles under positive aeration
- Curing

As noted above, more published data are needed on the use of covers. Their use will depend on location and regulation. Under certain conditions, the cost trade-off is between covers or enclosures, and covers are less expensive.

CHEMICAL ODOR NEUTRALIZERS AND ADDITIVES

There is also very little published information on the use of odor neutralizers in composting operations. Haug (1993) reviewed earlier effects of masking agents and neutralization agents on odor production. He indicated that neutrilizing agents were avaialble to counteract odor types such as amines, mercaptans, aldehydes, and aromatics.

Gruber et al. (1990) reported an average 44% reduction in effective dose at the 50% level (ED_{50}) after atomizing a neutralizing agent into the blower exhaust at a sludge composting facility. Other works were reported by Duffee (1988) and Canzano and Aiani (1988). These authors reported odor reduction ranging from 20 to 80%.

In a CIWMB (2008) study, two GOC™ Technologies commercial chemicals were added to source-separated, ground greenwaste. One additive was incorporated by turning the windrow following application. The second additive was applied topically. During the first week, there was a 42% VOC reduction with the chemical additives in comparison with the standard greenwaste windrow. However, after the first turning event on the seventh day, the effectiveness diminished. Over a fourteen-day period, the additive windrow generated 14% less VOCs than the standard greenhouse windrow.

Thus, it appears that chemical neutrilizing agents may be effective in reducing odors during composting operations. The value of these chemicals will depend on the technology. They may be more effective in static systems than in turning systems. In windrow systems, the cost of application following every turning could be expensive.

One of the additives that has been shown to be effective in reducing odors is wood ash (Carpenter and Beecher, 1997). Although wood ash has been used for several years by some facilities, it has not been definitively documented. In a study by the California Integrated Waste Management Board (CIWMB, 2002), wood ash was found to increase ammonia emissions on days 1 and 7 compared to the control. Ammonia generally disipates rapidly and is not offensive at a distance. This was believed to be the result of the strongly alkaline pH of the wood ash. The 25% wood ash treatment resulted in more effective treatment of odors for a longer period of time than lower wood ash additions. The study concluded that high-carbon wood ash can reduce odor and odorant emissions at composting facilities. Wood wastes may contain treated wood such as CCA, and therefore be rich in arsenic (As). The ash may contain as much as 30,000 ppm As. This may result in a high contamination of the compost and exceed state regulatory concentration levels.

CONCLUSION

Odor control during operation of composting facilities is extremely important. More facilities have been closed as a result of the production of odors offensive to residents nearby.

Odor management should begin with the design and establishment of the composting facility. Site conditions, climate or microclimate, location with respect to receptors, and the selection of the appropriate system and technology all can affect the release, dispersion, and impact of odors.

Operating composting facilities in relation to climatic and meteorological conditions can greatly reduce impacts to surrounding areas. Since during windrow turning odor emissions are the greatest, windrow turning should be done in relation to the climate.

Depending on the system used, there are numerous opportunities for managing and controlling odors.

Open systems such as windrows are less amenable to containing and treating odors. In those systems, as well as in many others, the management of odors is essential. This can consist of minimizing the release of odors during operation, when climatic conditions can result in impact to receptors; reducing odor emissions during turning or other operations; using fabric covers; and using sprays during turning. Using moisture sprays during windrow composting reduces odors. In addition, the possibility of using sprays containing additives or odor neutralizers can reduce odors. Addition of wood ash has been shown to reduce odors as long as there is no increase in arsenic content.

It is easier to control odors under negative air in ASP systems. The exhaust can be treated in a biofilter. The use of finished compost as a mini-biofilter or blanket cover is also effective in reducing odors and VOCs. Under positive air, the most effective way is the use of covers.

Where the odor can be contained and treated, biofilters have been shown to be the most cost-effective means of odor control. Biofiltration can be used for treating odors from enclosures used for mixing the feedstock with bulking agents and feedstock delivery. In within-vessel systems, such as agitated bed or others, biofiltration is a very effective means of odor control. ASP systems under negative aeration can also use biofiltration to control odors. The use of fabric covers in ASP systems has also been shown to be effective in odor control. Pseudobiofilters such as compost covers can be very effective in ASP and other static systems.

The use of covers can significantly reduce odors. Machines exist today that can place a cover over the windrow and remove it prior to turning. ASP systems can also benefit from the use of covers. Since it has been shown that curing odors can produce over 50% of the odors, covers can be effective to reduce these odors.

It is important to avoid site odors from ponding, spilled feedstocks, runoff, and other possible sources. These are easy to take care of and impress visitors that efforts are being made to have a well-run and maintained site.

Public relations is also important. It can help the operator identify the major source of annoying odors so that improvements can be made in that specific area. Further public relations indicate to the community that every attempt is being made to correct an unpleasant situation.

REFERENCES

ADL. 1989. *Odor evaluation program: Islip yard waste composting site*. Draft Report to Town of Islip Department of Environmental Protection, Ref. 64234. Cambridge, MA: Arthur D. Little, Inc.

Brenk, C., and Garibay, M. 1995. *Emission rate characterization of open windrow sludge composting operations*. Final report. Diamond Bar, CA: South Coast Air Quality Management District.

Buckner, S. C. 1996. *Aspects of process management and odor control during the composting of grass and other yard wastes*. Islip, NY: Department of Environmental Control.

Buckner, S. C. 2002. Effects of turning frequency and mixture composition on process conditions and odor concentrations during grass composting. In *2000 International Symposium Composting and Compost Utilization*, ed. J. F. C. Michel, R. F. Rynk, and H. A. Hoitink. Columbus: Ohio State University.

Buyuksonmez, F. 2007. *Comprehensive compost odor response project*. Prepared for the California Integrated Waste Management Board. Sacramento, CA: San Diego University.

Buyuksonmez, F., and Yucel, A. 2006. Odor mitigation alternatives. Paper presented at 14th Annual U.S. Composting Council Conference, Albuquerque, NM.

Canzano, P. S., and Aiani, K. J. 1988. Optimum system for odor control. *BioCycle* 29(7):40–44.

Carpenter, A., and Beecher, N. 1997. Wood ash finds niche in biosolids composting. *BioCycle* 38:37–39.

CH2MHill. 2005. *Cedar Grove composting Gore™ cover system testing pile no. 7 days 7, 14, 21, and 28 ammonia and VOC testing*. Everett, WA: CH2MHill.

CIWMB. 2002. *Using high-carbon wood ash to control odor at a green materials composting facility*. Sacramento: California Integrated Waste Management Board.

CIWMB. 2008. *Emissions testing of volatile organic compounds from greenwaste composting at the Modesto Compost Facility in the San Joaquin Valley*. Sacramento: California Integrated Waste Management Board.

Das, K. C. 2000. Odor related issues in commercial composting. Paper presented at the Workshop at the Y2K Composting in the Southeast Conference and Exhibition, Charlottesville, VA.

Duffee, R. A. 1988. *Deamine odor tests results relative to composting*. Report to NuTech Chemical Corp. Odor Science & Engineering, Inc., Bloomfield, CT.

Epstein, E. 1997. The science of composting. Boca Raton, FL: CRC Press LLC.

Epstein, E., Willson, G. B., Burge, W. D., Mullen, D. C., and Enkiri, N. K. 1976. A forced aeration system for composting wastewater sludge. *J. Water Pollut. Control Fed.* 48:688–94.

Epstein, E., and Wu, N. 2000. Planning, design, and operational factors that affect odor control at composting facilities. Paper presented at Y2K Composting in the Southeast, Charlottesville, VA.

Gruber, E. K., Harding, M. L., and Kuchenrither, R. D. 1990. Odor control at Philadelphia's sludge processing and distribution center. Paper presented at the Proceedings of the WPCF Specialty Conference Series, Status of Municipal Sludge Management, New Orleans, LA.

Haug, R. T. 1993. *The practical handbook of compost engineering*. Boca Raton, FL: Lewis Publishers.

Iacoboni, M. D., Leburn, T. J., and Lingstrom, J. 1980. Deep windrow composting of dewatered sewage sludge. In *Proceedings of the National Conference Municipal and Industrial Sludge Composting*, Philadelphia, PA, 88–109.

MOR. 2007. *Odor evaluation report*. Report 717902, prepared by St. Croix Sensory, Inc. for Managing Organic Recycling, Inc. San Louis Obispo, CA: Managing Organic Recycling.

MOR. 2009. *Compost odor study for Timpanogos Special Service District*. San Luis Obispo, CA: Managing Organic Recycling.

Murray, C. M., and Thompson, J. L. 1986. Strategies for aeration pile systems. *BioCycle* 27(7):22–28.

Nelson, D. W., and Sommers, L. E. 1982. Total carbon, organic carbon, and organic matter. In *Method of soil analysis*, ed. A. L. Page. Part 2. Madison, WS: American Society of Agronomy.

OS&E. 2005. *Evaluation of Gore™ cover system with a semi-permeable e-PTFE membrane technology for control of odors from greenwaste and biosolids composting*. Bloomfield, CT: Odor Science & Engineering, Inc.

Paula, E. A., Morris, S. J., Conanta, R. T., and Plantea, A. F. 2006. Does the acid hydrolysis-incubation method measure meaningful soil organic carbon pools? *Soil Sci. Soc. Am. J.* 70:1023–35.

Rynk, R., van de Kamp, M., Willson, G. B., Singley, M. E., Richard, T. L., Kolega, J. L., Gouin, F. R., Laliberty, L., Kay, D., Murphy, D. W., Hoitink, H. A. J., and Briton, W. F. 1992. *On-farm composting handbook*. Ithaca, NY: Northeast Regional Agricultural Engineering Service.

Schmidt, C. E., Card, T. R., and Kiehl, B. 2009. Composting trials evaluate VOC emissions control. *BioCycle* 50:33–36.

Tchobanoglous, G., Theisen, H., and Vigil, S. 1993. *Integrated solid waste management*. New York: McGraw-Hill.

TRC. 1991. *An analysis of odors from a yard-waste composting facility*. Seattle, WA: TRC Environmental Consultants.

11 Odor Control Systems

INTRODUCTION

The control of odors is paramount for the establishment and continued existence of a composting facility. More facilities have been shut down or engaged in litigation because of odors. As indicated earlier, odor management and control should be an important component of the planning, design, and establishment of a facility. Facility management is an important part of controlling odors. Odor management should begin with the design and establishment of the composting facility. Site conditions, climate or microclimate, location with respect to receptors, and the selection of the appropriate system and technology can all impact the release, dispersion, and impact of odors.

Regardless if one builds an odor control system, the site management to control odors is still applicable. There are many areas within a composting facility that normally are not included in an odor control system that could produce unacceptable odors.

The predominant odor control system used in composting is biofiltration, and it is applied to enclosures or conditions where the air can be contained and treated. Chemical systems are rarely used. In fact, in one facility that changed from a poor biological system to an expensive chemical system, off-site odors were worse and complaints more numerous. The control of odors during windrow operations is primarily through management.

The principal volatile organic compounds producing odors that need to be treated are sulfur and nitrogen compounds, as indicated in Chapter 9.

Public relations is also important. This can help the operator identify the major source and occurrence of anoying odors so that improvements can be made in that specific area. One facility brought neighbors who complained about odors to the site. They stated that it was not the odor smelt. It turned out that the annoying odor came from a farm. In another case, neighbors identified the odor as coming from a rendering plant and not the composting site. Furthermore, public relations indicates to the community that every attempt is being made to correct an unpleasant situation.

The objective of this chapter is to provide the reader with details and references to biofiltration and other odor control options.

BIOFILTRATION

PRINCIPLES OF BIOFILTRATION

Biofilters depend on microbial degradation of odorous compounds. As the odorous compound is passed through the media, it is adsorbed on the surfaces of the media particles or absorbed into the water film surrounding the media particles.

Microorganisms, principally bacteria but also actinomycetes and fungi, are attached to the filtering media (Williams and Miller, 1993). Heterotrophic bacteria utilize organic compounds and are the most responsible for removal of volatile organic compounds (VOCs). Some of the organisms involved in the biodegradation of organic compounds are *Pseudomonas*, *Actinomycete*, and *Norcadia*. There are two mechanisms involved simultaneously in the removal of odorous compounds: absorption/adsorption and biooxidation (Naylor et al., 1988). Allen and Hartenstein (1986) indicate that the uptake of the gas molecules by bacteria may occur in two ways. A gas molecule may diffuse to the surface of the filter media where it is adsorbed. It then is desorbed and taken up by a passing water droplet. In the second case, the gas molecule diffuses directly to the surface of a water film and is transferred across the gas-liquid interface. This is the reason for moisture being a key to a properly functioning biofilter, and explains why the efficiency decreases with decreasing humidity.

Waste gases containing odorous compounds and volatile organic compounds present in the exhaust gas from the composting operation are soluble in water and adsorbed into a moist biofilm in the biofilter media. This concept is illustrated in Figure 11.1.

Once inside the biofilm (microbial biomass and biofilm), microorganisms consume and degrade the compounds that are a source of energy to them. Several conditions are necessary for proper function of the biofilter. These are:

Water
pH and nutrients
Temperatures conducive to microbial growth
No toxic gases to microorganisms, such as high ammonia levels

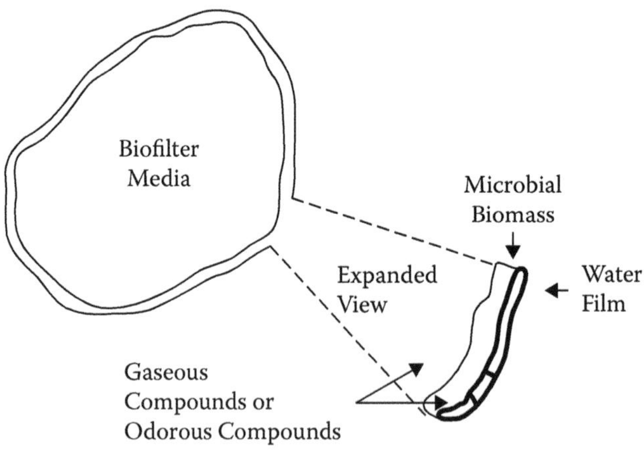

FIGURE 11.1 Conceptual model for adsorption and microbial degradation of gaseous odorous compounds. (Adapted from Eitner, D. *Investigations of the Use and Ability of Compost Filters for the Biological Waste Gas and Ability of Compost Filters for Biological Waste Gas Purification with Special Emphasis on the Operation Time Aspects* (in German), GWA, Aachen, West Germany, 1984.)

Water

Moisture is essential for microbial activity and for sustaining the proper microbial population. Too low a moisture content will result in decreased microbial activity. An excess of water blocks the pore space, reduces porosity or void space, and reduces the amount of oxygen available to microorganisms. Excess water can also result in greater head loss. The required amount of water varies with the media (Prokop and Bohn, 1985; Bohn and Bohn, 1988). Moisture is added through humidification of the exhaust gases and through water addition to the biofilter by the use of an irrigation system.

pH and Nutrients

Acidity (pH) affects the growth of microorganisms and their activity. Since organisms thrive at different pH levels, a near-neutral pH is desirable. When the predominant gas to be treated is hydrogen sulfide, sulfuric acid is formed, lowering the pH. To counteract this, lime or other sources of calcium carbonate (shells) have been used.

Normally there are sufficient inorganic and organic nutrients available for the growth and maintenance of the microbial population. Rarely are nutrients added.

Temperature

Since microorganisms are sensitive to temperature and can be destroyed at high temperatures, it is best to keep the temperature between 10 and 40°C (50 and 104°F) (Williams and Miller, 1993; Chiumenti et al., 2005). The latter authors indicate that above 40°C (100°F), there is lower biofiltration efficiency because of a decrease in mesophilic bacteria activity.

Toxic Gases

Ammonia is the most noted toxic gas to the microorganisms during composting that can affect the biofilter's performance. Ammonia levels above 100 ppm in the gas stream entering the biofilter can be toxic to microorganisms in the biofilter.

Advantages to Biofiltration

Cost—lower than that of other technologies
Reduced treatment residues—no disposal of residue or chemicals needed
Low operation costs
Low maintenance costs
High efficiency for compost odors and VOCs
Community acceptance—aesthetically acceptable

Disadvantages to Biofiltration

Space—uses more space than chemical systems
Temperature sensitive
Ammonia sensitive—levels of >100 ppm can be toxic
Changes and decrease in pressure across the bed, resulting in media removal and replacement

BIOFILTERS

The early biofilters used soil beds and were used to control odors from wastewater treatment plants and industrial plants (Bohn, 1975; Carlson and Leiser, 1982; Pomeroy, 1982). Allen and Hartenstein (1986) cite an early literature review of biofiltration. Bohn (1975) also reported on compost filters. County Sanitation Districts of Orange County, California, conducted an extensive review of the literature in 1993 (CSDOC, 1993). The USDA carried out the first use of compost biofilters to treat malodors from composting facilities in Beltsville, Maryland, during 1994 to 1997. Today biofilters are the most common odor control system at composting facilities. The efficiency of biofilters in composting operations has been well documented. The efficiency is highly dependent on physical and biochemical parameters. These include flow rate per unit area, bed temperature, filter media, moisture content, porosity, pH, loading rate, retention time, and other factors. Odor removal in some of the earlier composting facilities is shown in Table 11.1.

In Table 11.1, with the exception of the Hamilton, Ohio, biofilter, the average odor removal efficiency was 81%. In the 1980s and 1990s, considerable knowledge was obtained on the design of biofilters. In Table 11.2, with more recent data, the average

TABLE 11.1
Data on Odor Removal at Several Composting Facilities

	D/T Inlet		D/T Outlet		Odor Removal %	
Facility	**Range**	**Ave.**	**Range**	**Ave.**	**Range**	**Ave.**
Akron, OH	53–338	180	12–85	47	55–85	74
Hamilton, OH	158–289	223	84–158	127	0–47	31
Hampton Roads, VA	—	1700	—	200	—	88
Montgomery County, MD	175	230	52–94	63	67–76	72
Schenectady, NY	480–860	660	110–200	150	70–83	77
Schenectady, NY (after improvements of odor control)	—	558	—	21	—	96

Source: Adapted from Williams, T. O. 1995. *BioCycle* 36:49–56.

TABLE 11.2
Recent Data on Odor Removal at Several Composting Facilities

Facility	Facility Type	Odor Concentration at Biofilter Inlet (D/T)	Odor Concentration at Biofilter Outlet (D/T)	Removal Efficiency (%)
Williamstown, MA	Biosolids composting	458	22	95
Davenport, IA	Biosolids composting	354	37	90
Davenport, IA	Biosolids composting	640	108	60
Lancaster, PA	Biosolids composting	126	19	85
Yarmouth, MA	Biosolids composting	226	11	95
Plymouth, NH	Biosolids composting	227	23	90
Hamilton, OH	Biosolids composting	635	19	97
Auburn, ME	Biosolids composting	338	29	91
Concord, NH	Biosolids composting	670	6	99
Severville, TN	Biosolids/MSW	462	16	97
Marlborough, MA	Biosolids/MSW Co-composting	340	28	92
Cobb County, GA	Biosolids/MSW Co-composting	60	8	87
Average		344	27	92

Source: Alix, C., personal communication, 2002.

odor removal efficiency was 92%. Rosenfeld (1999) reported that the biofilter at the LRI Compost Factory was very efficient in removing light VOCs. Data reported by Kuter (1990) showed that odor removal by biofilters ranged from 90 to 100% for numerous compounds.

When malodorous volatile organic compounds are produced during composting, the compounds will vary with the feedstock and management of the composting operations. Composting biosolids emit different compounds than food waste, manure, or yard waste composting. The most common are hydrogen sulfide (H_2S), dimethyl sulfide, dimethyl disulfide, acetone, limonene, alpha pinene, methyl ethyl ketone, toluene, fluorotrichloromethane, and amines (van Durme et al., 1990; Lees and Tockman, 1987; Peterson et al., 2000).

TABLE 11.3
Biofilter Efficiency in Removing VOCs Produced during Composting

Compound	Biofilter Inlet ppb	Biofilter Outlet ppb	Percent Removal
Acetone	2,450	9.7	99.9
Carbon disulfide	17	5.1	70
2-Butanone	545	Not detected	100
Toluene	9.3	1.9	80
2-Hexanol	5.5	Not detected	100
Styrene	4.6	Not detected	100
m,p-Xylene	1.8	Trace	99
1,4-Dichlorobenzene	1.4	Trace	99
Methylene chloride	Not detected	Trace	—
Carbonyl sulfide	47	3.4	99
Methyl mercaptan	550	Not detected	100
Carbon disulfide	26	5.5	79
Dimethyl sulfide	294	Not detected	100
Dimethyl disulfide	266	Not detected	100
Hydrogen sulfide	60	Not detected	100

Source: Data based on a survey by E&A Environmental Consultants, Inc., 1994.

One of the major concerns by air quality agencies is the emission of VOCs. Table 11.3 shows some data on the removal of VOCs and sulfur compounds by a biofilter.

Kuter (1990) summarized removal rates for various odorous compounds by biofiltration. These are shown in Table 11.4. The data showed better than 90% removal of compounds.

DESIGN CRITERIA AND OPERATIONAL CONSIDERATIONS OF A BIOFILTER

Media Selection

The characteristics and properties of the media are important in the effectiveness of the biofilter. As was indicated earlier, several different media have been used. In the early stages of development, soil filters have been used (Bohn, 1975; Bohn and Bohn, 1988). Essentially, these are not used in composting operations today, as they take considerable space and the retention characteristics are poor. The most common media have been a combination of compost with wood chips; compost with wood chips and rubber tires; compost, bark, and soil; and peat. Several vendors offer a synthetic media, which can combine different components. Lava bedrock has also been used. Currently, a compost mixture is primarily used, as it is readily available and has been shown to be an excellent medium. The media must be able to maintain porosity for proper airflow. Therefore, some facilities have mixed wood chips or rubber tires with the compost. The media must provide a

TABLE 11.4
Removal of Several Odorous Compounds by Biofiltration

Odorous Compound	Removal Rate	Reference
Aldehydes	92–99%	Prokop and Bohn, 1985
Amines, amides	92–99%	Prokop and Bohn, 1985
Ammonia	92–95%	VDI, 1984
Benzene	>92%	Allen and Hartenstein, 1986
Cadaverine, putrescine, limonene	96%	VDI, 1984
Carbon monoxide	90%	Bohn and Bohn, 1988
Dimethyl sulfide	91%	Allen and Hartenstein, 1986
Ethanol, diacertyl, acetone	96%	VDI, 1984
Hydrogen sulfide	98–100%	Selby, 1986; Allen and Hartenstein, 1986
Isobutane, *n*-butane	95–98%	Bohn and Bohn, 1988
Mercaptans	92–95%	VDI, 1984
Organic acids	99.9%	Prokop and Bohn, 1985
Organic sulfides, disulfides	90–99.9%	Allen and Hartenstein, 1986; Prokop and Bohn, 1985
Polyaromatic hydrocarbons	95–100%	Bohn and Bohn, 1988
Propane	92–98%	Ebinger et al., 1987; Kampbell et al., 1987
Sulfur dioxide	97–99%	Bohn and Bohn, 1988
Terpenes	>98%	Allen and Hartenstein, 1986

Source: Kuter, G. A. *Odor Control, Completing the Composting Process*, Monograph, International Process Systems, Inc., Glastonbury, CT, 1990. With permission.

suitable environment for the microorganisms. It should be capable of adsorbing water, provide an air-filled pore space between 40 and 80%, have an organic matter content of 35 to 55%, and have a particle size distribution of more than 60% by weight of greater than 4 mm (0.157 inches) (Williams, 1994). The selection of the biofilter media can influence:

- Moisture control
- Media stability
- pH (buffering)
- Pressure drop
- Retention time
- Microorganisms

An example of a specification for a biofilter media is as follows:

Wood chips:-Wood chips shall consist of the large fraction of ground wood that has been screened through a 1.9 cm (0.75 inch) screen to remove the fine material. The wood chips shall be clean and free of nonwood material, such as plastic, glass, rock, and metal, and shall contain less than 4% bark.

The wood chips shall meet the following gradation requirements:

Sieve Size	Percent Passing
4 inch	90–100%
3 inch	60–90%
2 inch	20–60%
<1 inch	10%

Yard waste or biosolids compost: The leaf or yard waste or biosolids compost shall be produced through thermophilic-forced aeration or the turned windrow-composting process. The compost feedstock shall consist of leaves, grass, ground brush, and tree trimmings. Compost shall not contain manure or industrial sludge. All compost shall be cured to the point that it is no longer self-heating. The compost shall be free of objectionable odor caused by contaminants or ongoing decomposition.

The compost shall meet the following specifications:

Total solids	35–55%
Volatile solids	>25%
pH	6.0–8.5
Soluble salts	<5.0 mS/cm
Stability CO^2 respiration test	
	<5.0 mg CO^2-C/g compost C/day
Stability Solvita™ test	Maturity index 6 or higher
Man-made inert or foreign material	<1%
Particle size	
100% passing 2.54 (1 in.) sieve	
85% passing 0.95 cm (0.375 in.) sieve	

The biofilter media shall consist of a blend of three parts wood chips and one part yard waste or biosolids compost blended by volume. The material shall be thoroughly wetted and completely blended before placement.

The contractor is cautioned that after blending and wetting, the volume of the yard waste or biosolids compost will be significantly reduced such that the volume of the blended media will essentially be the same as the original volume of wood chips.

The blending shall be performed in the engineer's presence. The wood chips and yard waste or biosolids compost shall be thoroughly mixed by means of a front-end loader, and water shall be added to thoroughly wet the mixture.

The following blending procedures shall be employed:

Place a bed of wood chips (use approximately two-thirds of the wood chips to be blended). Add the compost to be blended to the wood chip bed. Add the remaining wood chips over the compost. With the front-end loader, lift the material to the full reach of the loader and allow the material to fall slowly by "feathering" the bucket. While the material is falling, spray a high volume of water through it. A minimum 1.3 cm (0.5 inch) hose is recommended. The blending shall be done until the mixture has the appearance of a mud-coated wood chip with no clean wood visible.

Placement of the biofilter media shall be done in the engineer's presence. The media to be placed shall consist of an initial 15 cm (6 inch) layer of wood chips over the biofilter air distribution floor, 1.2 m (4 feet) of blended media over the wood chips, and then another 15 cm (6 inch) layer of wood chips over the blended media.

> Placement of all layers may be done by machine, provided that the equipment will not exert a pressure greater than 2.91 kg/cm^2 (40 lb per square inch) on the biofilter bottom. Material must be placed by working in a small area at a time and placing all layers in that small area before moving on. At no time is any equipment allowed on any layer. No compacting of material on the surface is allowed.
>
> A smoke test is done to ensure that there is a uniform air distribution throughout the media. The smoke test is conducted at the following times: (1) after completion of the placement of the aeration plenum, but before wood chip or media placement, and (2) after media and wood chip layers have been placed.
>
> Prior to the initial smoke test, the biofilter air distribution channel and floor beneath the plenum shall be flushed clean, and all of the air distribution system shall be inspected to ensure clear openings. The smoke test cannot be performed until the facility ventilation system is completely operable. The facility ventilation system must be used for the test. The use of temporary fans for the test is not permitted. The contractor shall supply enough smoke bombs or other smoke generation equipment to smoke test the air distribution system for a minimum of five minutes to demonstrate the system's capability of even air distribution. If the test should fail to show even air distribution, the contractor shall make the necessary adjustment or repair to the system, and retest the system until satisfactory results are obtained. The smoke test to be performed after media placement shall also be for a duration of at least five minutes.

Media or bed depth serves to equalize the resistance to airflow, thus providing more uniform distribution of the incoming gas stream to the bed (Ostojic et al., 1989). Generally, the media depth recommended is 90 to 122 cm (3 to 4 feet) (Williams and Miller, 1993; Chiumenti et al., 2005). Deeper biofilters have been used to minimize the space requirement.

Moisture

Moisture content and its control are critical for the proper operation of a biofilter, since a water film is necessary for the absorption and biooxidation of the odorous compounds. Kuter (1990) recommended a moisture range between 40 to 60%. Haug (1993) recommends a moisture control of 50 to 70% by weight. Humidity must be maintained at greater than 99% (Bohn and Bohn, 1988). Moisture is typically controlled using sprinklers and by ensuring that the gas stream entering the biofilter is humidified.

Moisture maintenance and addition can be accomplished in two ways. The air entering the biofilter is humidified. The most common method is through a humidity chamber in advance of the biofilter. Alternatively, embedding very fine spray nozzles into the duct will also increase the humidity. In addition, it is recommended that an irrigation system be installed above the biofilter. This can be accomplished with soaker hoses or surface sprays. The amount of water needed depends on the climatic conditions of rainfall, temperature, and humidity. Runoff water can be used, but would most likely require a filter to avoid clogging of the system.

pH

A pH of between 7 and 8 is optimal. pH can be maintained or adjusted using lime or crushed shells.

Porosity

Porosity is the ratio of the pore space to the total volume of the media. Porosity affects the head loss through the biofilter. As the organic media degrades, small particles are produced that tend to plug up the media. Proper porosity also provides for the effective surface area for microbial activity. Excessive growth of biomass can clog the media.

In the example shown above, the ratio of wood chips to compost was 3:1. Sadaka et al. (2002) recommended a ratio of 80:20. The porosity reported ranged from 48 to 62%. The porosity of an organic medium changes since compost degrades with time. This results in the formation of fine particles, which will restrict airflow. Continual monitoring of the pressure drop across the filter will indicate short-circuiting or compaction (Williams and Miller, 1993). The effect of porosity (voids) is shown in Figure 11.2. The recommended porosity is between 50 and 55%.

Pressure Drop

The media bed porosity affects the efficiency of the biofilter. Pressure changes affect airflow. This is one reason that compost alone, even though it has been used, is usually blended with wood chips. The pressure drop across the biofilter should be noted. If there is a sudden drop, it usually indicates compaction. This may require changing the media or fluffing the bed. Plastic or shredded rubber tires have also been used. Pressure drop is increased by fine particles and compaction.

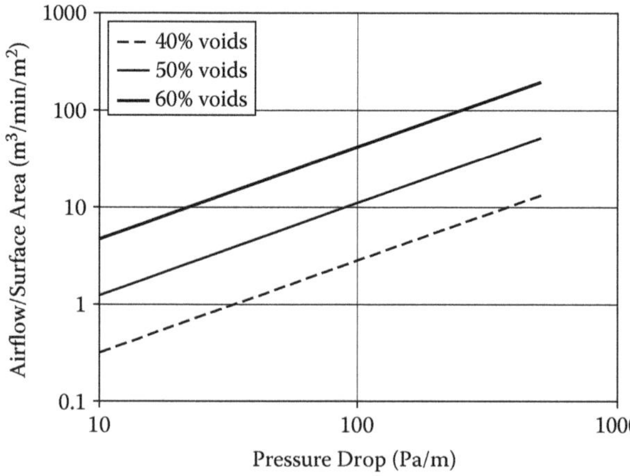

FIGURE 11.2 Effect of porosity (voids) on pressure drop. (After Nicolai, R. E. and Janni, K. A. Determining Pressure Drop through Compost-Woodchip Biofilter Media, paper presented at the ASAE Annual International Meeting, Sacramento, CA, 2001.)

Loading Rate

The loading rate is the rate of gaseous intake per unit area of a biofilter. The recommended rate is usually 1.21 to 1.52 m^3/m^2 (4 to 5 cfm/ft^2) of media area. Haug (1993) recommended a rate of (<100 $m^3/h\text{-}m^2$ (<5.4 cfm/ft^2)). Williams and Miller (1993) recommended a range of 0.3 to 9.5 $m^3/min/m^2$ (1 to 30 cfm/ft^2).

Temperature

It is recommended that the maximum inlet temperature to the biofilter be less than 40°C (104°F). In one study, VOC removal was reduced to 75% at 43°C (107°F). Since temperatures in compost piles can exceed 40°C (104°F), it may be necessary to dilute the air with outside air. Often installation of sprinkler nozzles with a very fine spray can be placed in the duct leading from the air source to the biofilter. This can serve two purposes by reducing temperature and reducing ammonia. The temperature in the bed should be in the range of 2 to 40°C (36 to 104°F) (Kuter, 1990). At higher temperatures microbial die-off occurs. Often the temperature of the exhaust gas exceeds 80°C (176°F) and needs to be lowered. This can be accomplished by providing moisture to the exhaust or by mixing with outdoor air.

Ammonia

The level of ammonia in the gas stream can be reduced by dilution with fresh air, using a dilute acid spray within the humidification chamber, or using a dilute acid spray in thc duct. When using a dilute acid, care must be given to corrosion.

Nutrients

Nutrients are not usually added, as they are in sufficient supply in the biofilter media.

Retention Time, Residence, or Detention Time

The recommended residence or detention time is thirty to sixty seconds. Gas retention time affects the removal efficiency of the odorous compounds. This is illustrated in Figure 11.3. Yang and Allen (1994) reported that H_2S removal was efficient as long as the retention time exceeded twenty-three seconds. Currently a detention time of forty-five seconds is being recommended.

BIOFILTER DESIGN

The design of the biofilter consists of the following elements:

- Blower
- Humidification system
- Air distribution system—channels/trenches, perforated blocks, perforated pipe
- Stone
- Media
- Irrigation system

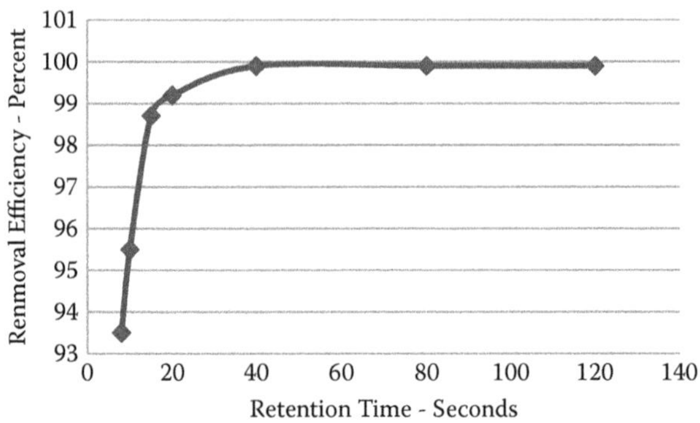

FIGURE 11.3 Effect of gas retention time on hydrogen sulfide removal efficiency. (Based on Yang, Y. and Allen, E. *J. Air Waste Manage. Assoc.*, 44, 863–68, 1994.)

Most biofilters are aboveground. However, there is no reason why a biofilter cannot be belowground. The Heidelberg, Germany, composting facility has a belowground biofilter. If a biofilter is belowground, there is the opportunity to have a cover over it to capture the emissions and treat them by other means if necessary and desirable.

Two basic illustrations are shown in Figures 11.4 and 11.5. In Figure 11.4 the key elements are the air-intake system to the biofilter, the humidification system, and the biofilter consisting of a distribution system and media. Figure 11.5 illustrates the distribution system and how the media adsorb and absorb foul air.

Pictured in Figure 11.6 is an illustration of a multicell biofilter design. The multicell biofilter captures emissions from different sources and uses a single biofilter to treat the emissions from these sources. For example, if the three sources are from different buildings, such as a feedstock preparation room, a composting hall, and a curing hall, a single biofilter can be designed having different size sections to treat the different volumes and concentrations of odors and emissions.

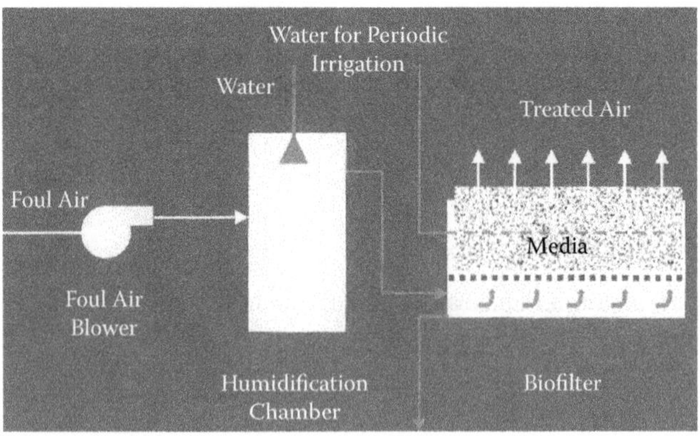

FIGURE 11.4 Generalized illustration of components of a biofilter. (From Groskreutz, R. Biofilters for Emission and Odor Control, paper presented at SCAP Biosolids Workshop, CA, 2004. With permission.)

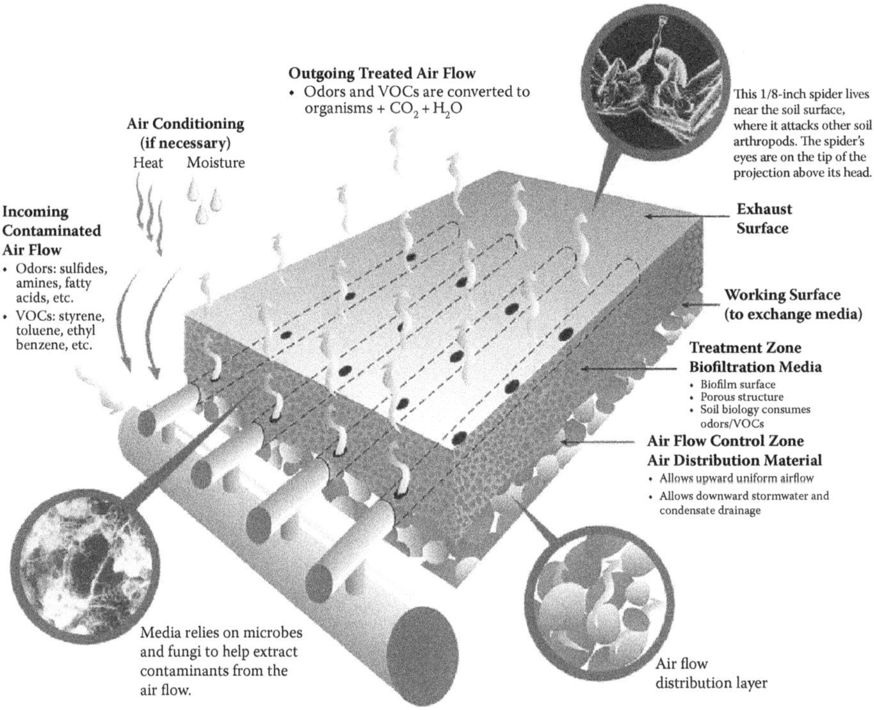

FIGURE 11.5 Schematic of biofilter showing functional elements. (After Groskreutz, R. Biofilters for Emission and Odor Control, paper presented at SCAP Biosolids Workshop, CA, 2004. With permission.)

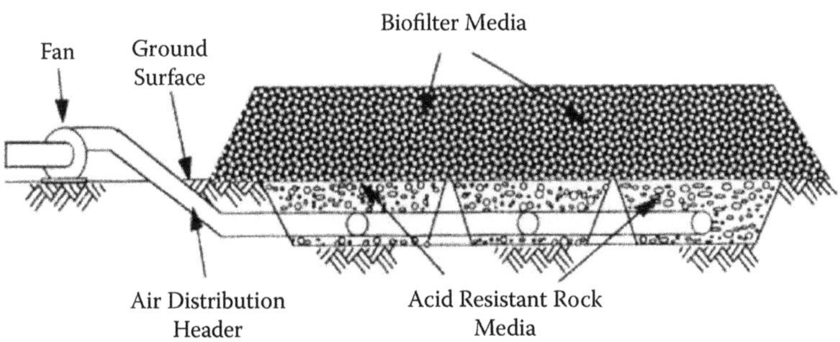

FIGURE 11.6 Multicell arrangement in a biofilter.

For a large biofilter, it is advisable to have an engineering firm with experience in designing biofilters.

Examples of biofilters in use are shown in Figures 11.7 to 11.9. In Figure 11.7 the biofilter treats the air from one-half of an ASP system for biosolids and yard waste located inside a building. The biofilter has been operating for fifteen years, and there

FIGURE 11.7 An aboveground biofilter in Davenport, Iowa.

FIGURE 11.8 Biofilter in the Wright system at Disney World, Florida.

FIGURE 11.9 Biofilter designed by the author at a Victoria, Canada, wastewater treatment plant during smoke testing.

have not been complaints from residents in a nearby community. The biofilter media has been changed several times.

In Figure 11.8 the biofilter treats the air from a food waste, within-vessel system designed by Wright and located in Disney World, Orlando, Florida. In this biofilter design, if additional capture of VOCs is needed, it would be quite easy.

Figure 11.9 shows a biofilter the author built at a small wastewater treatment plant. Since limited space was available to build the biofilter, it was built on part of a parking lot. For over ten years residents complained of odors. An odor survey and discussions with the residents determined that the odors were primarily sulfur compounds, which could be easily captured by a biofilter. A vendor-supplied biofilter is shown in Figure 11.10. Numerous vendors provide biofilters. The following are just a few: Bioton, Bdt BiDigestor Technologies, Bay Products, BioCube, Bio Reactor, Biorem, Envirogen, and Zabocs.

MANAGING BIOFILTERS

The following are the major variables that affect biofilter performance:

- The condition of the biofilter surface. It is important to prevent plant growth. The root systems of vegetation growing on the biofilter surface provide a pathway for the air flowing through the biofilter to result in short-circuiting. Excessive growth in dense patches of groundcover plants cause blocking of airflow and disruption of the loading rate throughout the biofilter.

FIGURE 11.10 A vendor-supplied biofilter at a composting facility in Hong Kong.

The best method of removal is hand weeding, which, if performed on a regular basis, is simple. Mechanical tilling has been used on biofilters but is not recommended because it speeds the decomposition process, reducing the life of the areas tilled. In addition, tilling equipment is heavy, and repeated use compacts the biofilter media. Biodegradable herbicides have also been used, especially with large biofilters.

- The moisture content of the media/moisture control through irrigation. Usually irrigation heads placed in corners or other sections of the biofilter can provide the necessary water. The optimum moisture content for the blended media is between 55 and 60%, with the biofilter continuing to work up to 50% moisture. To maintain this moisture level in the biofilter, water must be added.
- The pH of the media. The pH of the media should not drop rapidly. Since the sulfur compounds in the air stream are low, it is advisable to periodically check the pH of the media. If a significant drop in pH occurs, then the media will have to be replaced.
- Head loss. Periodic measurement of head loss is necessary. It is advisable to check the head pressure across the fans. Head loss should be expected to gradually increase over time. However, if large changes (increases or decreases) occur, there may be problems. Increases in back pressure can be a sign of an obstruction in the ductwork or dust clogging the holes of the distribution piping in the biofilter. Pressure increases can also be caused by water building up in the stone plenum layer to a level above the holes in the distribution piping; this would be caused by a blockage in the leachate

drain system. Sudden drops in the back pressure mean that channeling may be occurring in the biofilter media. Channeling is evidenced by cracking or excessively dry areas of the media.

OTHER BIOLOGICAL CONTROL SYSTEMS

Bioscrubbers can be used where space is limited. There are two types: suspended-growth bioscrubbers (Figure 11.11), in which microorganisms are suspended in a liquid, and fixed-film bioscrubbers, in which microorganisms are attached to a packing material. Figure 11.12 shows a combination of a biofilter and a biotrickling filter for maximum odor and VOC removal.

Biotrickling filters use synthetic media where bacteria decontaminate VOCs and odorous compounds. In both bioscrubbers and biotrickling filters, better controls can be had in contrast to biofilters. Bioscrubbers and biotrickling filters are rarely used in composting facilities. In the past Cape May, New Jersey, used a packed-bed bioscrubber. These systems are rarely used in composting facilities.

COVERS

Recently, the potential use of several different fabric covers has been investigated to reduce odors. The efficiency of covers was discussed in Chapter 10. During the past few years, there has been considerable improvement in cover efficiency as to reduction of emissions of both VOCs and odors. This is the result in modifications in fabric porosity. These recent covers mitigate odors and VOCs very effectively. There is very limited published data on the efficiency of fabric covers to reduce odors and emissions. Most of the data are available through reports and are not peer reviewed. This does not diminish the value of the data. Currently, there are two major companies

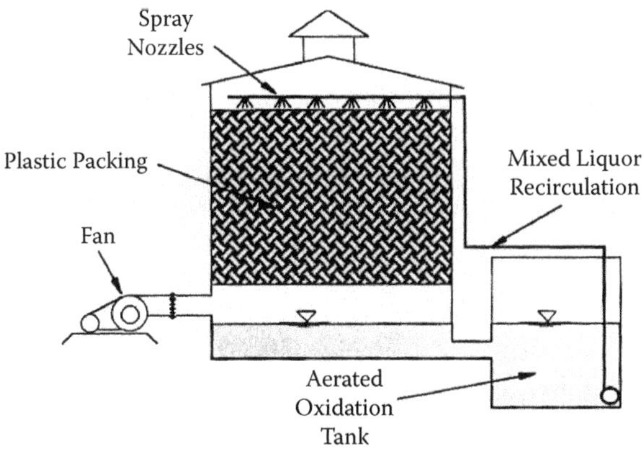

FIGURE 11.11 Typical suspended growth bioscrubber schematic. (From Mahin, T., ed., *Control of Odors and Emissions from Wastewater Treatment Plants*, WEF Manual of Practice 25, Water Environment Federation, Alexandria, VA, 2004. With permission.)

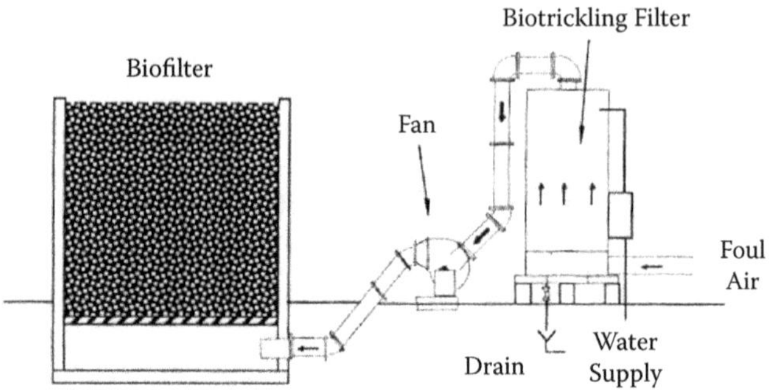

FIGURE 11.12 Combination of a biotrickling filter and a biofilter. (From Mahin, T., ed., *Control of Odors and Emissions from Wastewater Treatment Plants*, WEF Manual of Practice 25, Water Environment Federation, Alexandria, VA, 2004. With permission.)

providing covers in the United States. One is Managed Organic Recycling (MOR) and the other Gore™. Both these companies have equipment that applies the covers and is able to remove them. Figures 11.13 and 11.14 show the MOR covers and equipment, and Figures 11.15 and Figure 11.16 show the Gore system and equipment.

Since equipment is now available to remove and replace covers over windrows, under certain conditons their use would be very appropriate and advisable.

Covers appear to be excellent in reducing odors and VOCs by over 75%. They are particularly effective under the following conditions:

- Windrows—machines are available for placing and removing covers
- Passive aerated piles
- Aerated static piles under positive aeration
- Curing

Their use will depend on location and regulation. The closer a facility to receptors, the greater odor controls that will be needed. When designing a facility and its operation, in addition to location and proximity to potential receptors, consideration needs to be given to the type and quantity of various feedstocks the facility will accept. Under certain conditions, the cost trade-off is between covers or enclosures, and covers are less expensive.

CONCLUSION

Odor control during operation of composting facilities is extremely important. Many facilitiess have been closed as a result of the production of odors offensive to residents nearby.

Odor management should begin with the design and establishment of the composting facility. Site conditions, climate or microclimate, location with respect to receptors, and the selection of the appropriate system and technology all can impact the release, dispersion, and impact of odors.

FIGURE 11.13 Managed Organic Recycling (MOR) fabric covers with aerations system in Central Valley, Utah.

FIGURE 11.14 Managed Organic Recycling (MOR) equipment used for placing and removing fabric covers. (Courtesy of Managed Organic Recycling, Inc.)

FIGURE 11.15 The Gore fabrics cover system at Cedar Grove composting facility in Washington state. (Courtesy of W.L. Gore & Associates, Inc.)

FIGURE 11.16 The Gore equipment for placing and removing fabric covers. (Courtesy of W.L. Gore & Associates, Inc.)

Where the odor can be contained and treated, biofilters have been shown to be the most cost-effective means of odor control. Biofiltration can be used for treating odors from enclosures used for mixing the feedstock with bulking agents and feedstock delivery. Within-vessel systems, such as agitated bed or others, biofiltration is a very effective measns of odor control. ASP systems under negative aeration can also use biofiltration to control odors.

The use of fabric covers in ASP systems has also been shown to be effective in odor control. Psuedobiofilters such as compost covers can be very effective in ASP and other static systems. This aspect is discussed in Chapter 10.

Open systems such as windrows are more difficult to contain and treat odors. In those systems, as well as in many others, the management of odors is essential. This can consist of minimizing the release of odors during operation, when climatic conditions can result in impact to receptors; reducing odor emissions during turning or other operations; using fabric covers; and using sprays during turning.

Public relations is also important. It can help the operator identify the major source of annoying odors so that improvements can be made in that specific area. Further public relations indicates to the community that every attempt is being made to correct an unpleasant situation.

REFERENCES

Allen, E. R., and Hartenstein, H. U. 1986. *Biofiltration an odor control technology for a wastewater treatment facility*. Gainesville, FL: Environmental Engineering Sciences Department, University of Florida.

Bohn, H. L. 1975. Soil and compost filters of malodorant gases. *J. Air Pollut. Control Assoc.* 25(9):953–55.

Bohn, H., and Bohn, R. 1988. Soil beds weed out air pollutants. *Chem. Eng.* 95(6):73–76.

Carlson, D. A., and Leiser, C. P. 1966. Soil beds for the control of sewage odors. *J. Poll. Control Fed.* 38(5):829–40.

Chiumenti, A., Chiumenti, R., Diaz, L., Savage, G. M., Eggerth, L. L., and Goldstein, N. 2005. *Modern composting technologies*. Emmaus, PA: BioCycle.

CSDOC. 1993. Biofiltration research project, literature review report. Fountain Valley, CA: County Sanitation Districts of Orange County.

Ebinger, M. H., Bohn, H., and Puls, R. W. 1987. Propane removal from propane-air mixtures by soil beds. *JAPCA* 37:1486–89.

Eitner, D. 1984. *Investigations of the use and ability of compost filters for the biological waste gas and ability of compost filters for biological waste gas purification with special emphasis on the operation time aspects* (in German). Aachen, West Germany: GWA.

Groskreutz, R. 2004. Biofilters for emission and odor control. Paper presented at SCAP Biosolids Workshop, CA.

Haug, R. T. 1993. *The practical handbook of compost engineering*. Boca Raton, FL: Lewis Publishers.

Kampbell, D. H., Wilson, J. T., Read, H. W., and Stocksdale, T. T. 1987. Removal of volatile aliphatic hydrocarbons in a soil bioreactor. *JAPCA* 37:1236–40.

Kuter, G. A. 1990. *Odor control, completing the composting process*. Monograph. Glastonbury, CT: International Process Systems, Inc.

Lees, P. S. J., and Tockman, M. S. 1987. *Evaluation of possible health impacts of WSSC Site II sewage sludge composting operations*. Report to Maryland Department of Health and Mental Hygiene. Baltimore: Johns Hopkins University, School of Hygiene and Public Health.

Mahin, T., ed. 2004. *Control of odors and emissions from wastewater treatment plants*. WEF Manual of Practice 25. Alexandria, VA: Water Environment Federation.

Naylor, L. M., Kuter, G. A., and Gormsen, P. J. 1988. *Biofilters for odor control: The scientific basis*. Hampton, NH: International Process Systems, Inc.

Nicolai, R. E., and Janni, K. A. 2001. Determining pressure drop through compost-woodchip biofilter media. Paper presented at the ASAE Annual International Meeting, Sacramento, CA.

Ostojic, N., Duffe, R. A., and O'Brien, M. A. 1989. Use of biofiltration for odor control. Paper presented at Air and Waste Management Association International Specialty Conference on Recent Developments in Odor Regulations, Controls, and Technology, Detroit, MI.

Peterson, M. K., Johnson, D. L., and Pleus, C. 2000. Characterization of emissions from two yard-waste composting facilities. In *Odors and VOC Emissions 2000*. Cincinnati, OH: Water Environment Federation.

Pomeroy, R. D. 1982. Biological treatment of odorous air. *J. Water Pollution Control Fed.* 54:1541–1545.

Prokop, W. H., and Bohn, H. L. 1985. Soil bed system for control of rendering plant odors. *J. Air Pollut. Control Assoc.* 35:1332–38.

Rosenfeld, P. 1999. *Characterization and quantification of odorants from the LRI Compost Factory*. Report to LRI. Seattle, WA.

Sadaka, S., Halbach, T. R., and Mann, D. D. 2002. Vertical and horizontal airflow characteristics of wood/compost mixtures. *Appl. Eng. Agric.* 18:735–41.

Selby, M. A. 1986. Soil filters at treatment plant: Potential for soil is utilized at Tamarac, Florida facility. *BioCycle* 27:33.

Van Durme, G. P., McNamara, B. F., and McGinley, V. 1990. Characterization and treatment of composting emissions at Hampton Roads Sanitation District. Paper presented at the 63rd Annual Water Pollution Control Federation Conference, Washington, DC.

VDI. 1984. *Biologische abluftreinigung—Biofilter* [Biological waste air putrification]. Handbuch Reinhaltung der Luft, West Germany: VDI.

Williams, T. O. 1994. Biofiltration for control of odorous emissions and volatile organic compounds (VOC's) from wastewater and sludge processing facilities. Paper presented at Water Environment Federation Specialty Conference Series on Odor and VOC Emission Control for Municipal and Industrial Wastewater Treatment Facilities, Jacksonville, FL.

Williams, T. O. 1995. Odors and VOC emissions control. *BioCycle* 36:49–56.

Williams, T. O., and Miller, F. C. 1993. Composting facility odor control using biofilters. In *Science and engineering of composting: Design, environmental, microbial, and utilization aspects*, ed. H. A. J. Hoitink and H. M. Keener. Wooster: Ohio State University.

Yang, Y., and Allen, E. 1994. Biofiltration control of hydrogen sulfide. 1. Design and operational parameters. *J. Air Waste Manage. Assoc.* 44:863–68.

12 Pathogens

INTRODUCTION

In addition to odors, two very important environmental composting management aspects need to be considered in the design and operation of composting facilities. These are:

- Pathogens
- Bioaerosols

The U.S. Environmental Protection Agency (USEPA) regulates pathogens for only one feedstock, sewage sludge. Most states apply these regulations to other feedstocks.

Most of the feedstocks currently being composted contain pathogens. It is a misconception that feedstocks such as yard waste do not contain pathogens, and therefore should not be required to meet the time-temperature regulations as suggested by states. Pathogens are a major concern to citizens residing near composting facilities, as well as a concern for the health of workers. Therefore, facilities need to be designed to minimize the dispersion of pathogens and to manage operations in a sound environmental manner.

The destruction of pathogens and the control of vectors that may transmit pathogens are paramount to a successful composting operation. There are several aspects to the management of facilities with respect to pathogen destruction and vector attraction reduction. The facility's operation can affect worker health within the facility, and dispersion from the facility can impact public health for individuals outside the facility. Furthermore, the destruction of pathogens is vitally important in the production of an acceptable product for marketing and utilization. A key to the marketing of compost is the production of a disinfected product. There is a need for the public to be confident that the product is safe and will not impair their health.

The key parameter for the destruction of pathogens is the temperature-time relationships (Epstein, 1997). This is the basis for the USEPA 40CFR503 regulations related to composting of sewage sludge and biosolids (USEPA, 2003). One issue is that of pathogen regrowth or recontamination, and subsequently increased numbers in a product. According to the USEPA, pathogen regrowth may occur unless (1) an inhibitory chemical is present, (2) the material is too dry to allow bacterial growth, (3) little food remains for the microorganisms to consume, or (4) an abundant population of nonpathogenic bacteria is present (USEPA, 1999). The USEPA 40CFR503, "Vector Attraction Reduction (VAR)," requirements are designed to provide the removal of sufficient pathogen food source by degradation and stability in the composting material so that pathogen regrowth is unlikely. These regulations were promulgated to disinfect sewage sludge and biosolids. There are no federal regulations pertaining to other waste materials. Many states, however, use the federal regulations for other wastes, such as solid waste, manure, food waste, and yard waste. The 40CFR503 regulations do not require further stabilization after VAR requirements are met. However, material meeting VAR requirements is not normally defined as stable according to the U.S. Composting Council (USCC) TMECC 5.08-B standards. Additional composting and curing further ensure pathogen reduction and are required to produce a USCC-defined stable and mature compost product. Research has shown that effective pathogen destruction occurs during curing. Stability requirements suggested in this chapter provide an additional level of protection from pathogen regrowth. Furthermore, enhancement of the indigenous bacterial population will further prevent pathogen regrowth.

Every aspect of our environment and our life involves pathogens. The three components of the environment—water, soil, and air—contain pathogens. Figure 12.1 illustrates the various pathogen pathways.

Most people who are not immunocompromised and are able to maintain health through proper diet, nutrition, and hygiene are generally in good health and are less impacted by pathogens in the environment. For an infectious disease to occur, there has to be an unsuccessful relationship between the host (human) and a pathogenic organism. Infection is initiated on body surfaces such as skin and mucous of membranes of the respiratory, gastrointestinal, and genitourinary tracts. This relationship is enhanced when our bodies are compromised, debilitated, and weak. An infection does not necessarily result in a disease.

How many organisms are needed to cause an infection or disease? The relationship relating the number of organisms that make individuals ill to carriers is termed infective dose. It varies with the organism or pathogen. Tolerance among individuals varies and is generally lower for the young, elderly, and immuncompromised (Bryan, 1977;

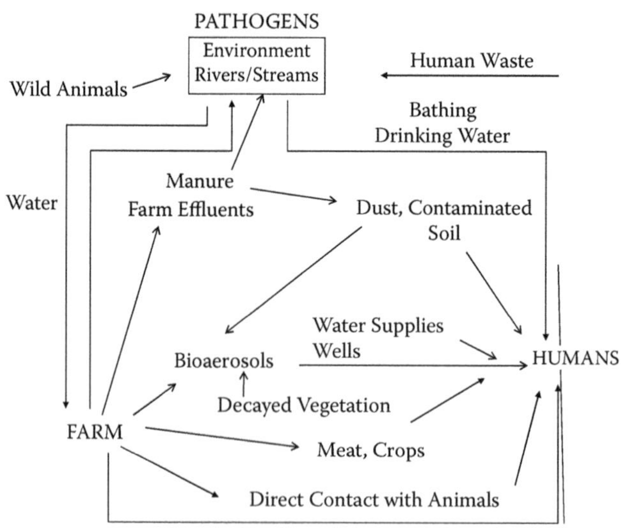

FIGURE 12.1 Pathogen pathways in the environment.

Cliver, 1980; Kowal, 1982). The data for humans are difficult to obtain. Akin (1983) reviewed the early literature on infective dose data for enteroviruses and other pathogens. The widest dose-response range occurred with enteric bacteria. *Salmonella* spp. required 10^5 to 10^8 cells to produce a 50% disease rate in healthy adults. Three species of *Shigella* produced illness in subjects administered ten to one hundred organisms. Administering small doses (one to ten) of cysts of *Entamoeba coli* and *Giardia lamblia* caused amoebic infections. Very low doses of enteric viruses were found to produce infection. Hornick et al. (1970) administered various doses of *Salmonella typhi* to adult volunteers and found that none of the fourteen volunteers showed any symptoms when one thousand organisms were administered. Twenty-eight percent of the adults became ill when a dose of one hundred thousand was administered, and 95% of the subjects were ill when 1 billion organisms were administered.

Table 12.1 shows the infective dose of various human pathogens. Most of the data are from the mid-1980s.

This chapter discusses environmental management with respect to primary pathogens. A subsequent chapter will discuss bioaerosols, which are often defined as secondary pathogens. Primary pathogens are those pathogens that can invade and infect humans regardless of their health status. Secondary pathogens are those pathogens that normally do not infect the health of humans but invade and infect debilitated, immunodeficient, or immunocompromised individuals. Introduced in this chapter is a discussion of current aspects of emerging pathogens. The term *emerging* does not necessarily imply new pathogens but includes pathogens that have become major health concerns.

There are several excellent resources:

Smith, J. E., Millner, P. D., Jakubowski, W., Goldstein, N., and Rynk, R., ed., *Contemporary Perspectives on Infectious Disease Agents in Sewage Sludge and Manure*, JG Press, Inc., Emmaus, PA, 2005

Bowman, D. D., ed., *Manure Pathogens: Manure Management, Regulations, and Water Quality Protection*, WEF Press, Alexandria, VA, 2009

Epstein, E., *The Science of Composting*, CRC Press, Boca Raton, FL, 1997

FDA, *Bad Bug Book, Foodborne Pathogenic Microorganisms and Natural Toxins Handbook*, FDA, Washington, DC, 2009

PATHOGEN CONCENTRATIONS AS RELATED TO FEEDSTOCKS

Most feedstocks can contain pathogens. Exceptions are industrial waste, such as pulp and paper mill sludge, where the sanitary sewage is separated from the sludge produced during processing. The following composting feedstocks may contain pathogens:

- Sewage sludge and septage
- Biosolids
- Municipal solid waste
- Food waste
- Manures
- Yard materials

TABLE 12.1
Infective Dose of Various Human Pathogens

Organism	Infective Dose	Range	Reference
		Bacteria	
Campylobacter spp.	100–500	100–500	Robinson, 1981; Cole et al., 1999
Clostridium perfringes	$>10^8$	10^6–10^{10}	Kowal, 1985; Brooks et al., 1991
Escherichia coli	10^4	10^4–10^{10}	Kowal, 1985; Keswick, 1984
E. coli O157:H7	50–500	50–500	Jones, 1999
Salmonella spp.	10^2	10^2–10^{10}	Kowal, 1985; Bitten, 1994
Shigella dysenteriae	10–10^2	10–10^9	Kowal, 1985; Keswick, 1984; Levine, 1973
Streptococcus faecalis	10^9	10^9–10^{10}	Kowal, 1985
Vibrio cholerae	10^3	10^3–10^{11}	Kowal, 1985; Keswick, 1984
		Viruses	
Echovirus 12	HID_{50} 919 PFU	17–919 PFU	Kowal, 1985
Poliovirus	$TCID_{50}$, <1 PFU	0.2–5.5 × 10^6 PFU for infants	Kowal, 1985
Rotavirus	$HID_{50,}$ 10 ffu	0.9–10^4 ffu	Ward et al., 1986
		Parasites	
Entamoeba coli	1–10 cysts	1–10 cysts	Kowal, 1985
Cryptosporidium	10 cysts	10–1,000 cysts	Robinson, 1981; Casemore, 1991; Okhuysen et al., 1999
Giardia lamblia	1 cyst estimated	NR	Kowal, 1985
Helminths	1 egg	Not reported	Kowal, 1985

Note: HID = human infective dose, $TCID_{50}$ = tissue culture infectious dose for 50% response, PFU = plaque-forming unit, ffu = focus-forming unit, NR = not reported.

Pathogens in Sewage Sludge, Septage, and Biosolids

Sewage sludge and septage contain pathogens, as illustrated in Table 12.2. Biosolids are treated sewage sludges. USEPA, in promulgating 40CFR503, indicated that pathogen destruction, with subsequent vector attraction reduction, was paramount if sewage sludge and septage were to be used for beneficial use. The early regulations required the use of technology-based methods for treatment of sewage sludge to reduce pathogens. Subsequently, direct enumeration of pathogens in biosolids was required.

Pathogens in Manures

Manures can contain pathogens that can infect humans and cause disease (Pell, 1997; Bicudo and Goyal, 2003; Hess et al., 2004). Hess et al. (2004) reported that

TABLE 12.2
Principal Pathogens Found in Sewage Sludge, Which Can Cause Diseases in Humans

Organism	Disease
Bacteria	
Salmonella spp.	Salmonellosis (food poisoning), typhoid
Shigella spp.	Dysentery
Yersinia spp.	Gastroenteritis
Vibrio cholerae	Cholera
Campylobacter jejuni	Gastroenteritis
Escherichia coli (pathogenic strains)	Gastroenteritis
Viruses	
Poliovirus	Poliomyelitis
Coxsackievirus	Meningitis, pneumonia, hepatitis, fever
Echovirus	Meningitis, paralysis, encephalitis, fever
Hepatitis A virus	Infectious hepatitis
Rotavirus	Gastroenteritis
Norwalk agents	Gastroenteritis
Reovirus	Respiratory infections, gastroenteritis
Parasites	
Protozoa	
Cryptosporidium parvum	Cryptosporidiosis
Entamoeba histolytica	Enteritis
Giardia lamblia	Giardiasis
Balantidium coli	Diarrhea and dysentery
Toxoplasma gondii	Toxoplasmosis
Helminth Worms	
Ascaris lumbricoides	Ascariasis, digestive disturbances
Ascaris suum	Coughing, chest pain
Trichuris trichiura	Abdominal pain, diarrhea, anemia
Toxocara canis	Fever, abdominal discomfort, neurological symptoms
Taenia saginata	Taeniasis, anorexia, insomnia, abdominal pain
Taenia solium	Taeniasis, anorexia, insomnia, anorexia
Necator americanus	Hookworm disease
Hymenolepis nana	Taeniasis

E. coli survived in outdoor raw manure for twenty-one months, and the concentration ranged from 10^2 to 10^6 CFU/g. Most manure that is applied to crops is not treated by composting, alkaline stabilization, or heat treatment. Reported outbreaks have been related to *Salmonella* spp., *Escherichia coli* O157:H7, *Cryptosporidium parvum, Mycobacterium avium* subspecies *paratuberculosis, Yersinia enterocolitica, Clostridium perfringens, Bacillus cerus, Enterococcus*, and *Listeria monocytogenes* (Valcour et al., 2002; Millner and Karnes, 2005). Derbyshire (1976) identified enteroviruses, adenoviruses, and a coronavirus from liquid swine manure. Human enteric viruses were detected in large numbers from water samples potentially contaminated by pig wastes (Payment, 1989).

Table 12.3 shows some of the pathogens in animal manures that can cause diseases in humans. There have been several incidences of food contaminated through the use of untreated (raw) manure. Several of these incidences were with the pathogen *E. coli* O157:H7. One of the greatest incidences because of drinking contaminated water was due to *Cryptosporidium parvum.* This incidence occurred in Milwaukee, Wisconsin, where 403,000 persons became ill and over 100 died (Mackenzie et al., 1994). It was thought to be because of water resources contamination by animal waste. Outbreaks of *Listeria monocytogenes* on raw vegetables have occurred because of fertilization with sheep manure (Pell, 1997). There has been concern that manure from a diseased animal could contain prions. Prions are associated with the nonpathogenic mad cow disease, which can infect humans. There is no evidence that animal manure contains prions.

Pathogens in Other Feedstocks

Solid waste contains pathogens. The major possible sources can be diapers, animal feces, and food waste (Pahren, 1987; Gerba et al., 1995; Gerba, 1996). Disposable diapers may contain enteric viruses that are excreted in the feces of infants and young children (Huber et al., 1994). Table 12.4 shows some data on the indicator organisms, total and fecal coliforms, and fecal streptococci. The fecal streptococcus group consists of a number of species of the genus *Streptococcus,* such as *S. faecalis, S. faecium, S. avium, S. bovis, S. equinus,* and *S. gallinarum.* These organisms are indicators of the presence of fecal matter from warm-blooded animals, including humans.

Data on pathogens in food waste, yard waste, and other feedstocks are very limited. Table 12.5 shows some early data. In a more recent study in Oregon on the composting of commercial food waste, the following data were obtained. Fecal coliforms ranged from <38,000 to 1.6×10^5 MPN/g total solids, and *Salmonella* spp. ranged from 100 to 510 MPN/4 g total solids (Epstein et al., 2005).

Contamination of our water resources and food can result in major waterborne and foodborne diseases. In the period 1991 to 2002, 207 waterborne disease outbreaks and 433,947 illnesses were reported (Craun et al., 2006). Foodborne illnesses are estimated at 76 million and result in 5,200 deaths in the United States each year (CDC, 2005).

In recent years several outbreaks of foodborne illnesses have occurred. A number of these incidences have been associated with domestic animal feces or animal manures. The microorganisms indicated were *Escherichia coli* O157:H7, and verotoxigenic, enterohemorrhagic strains, *Salmonella typhi*, *Yersinia entercololitica*

TABLE 12.3
Various Pathogens in Animal Manures That Can Cause Diseases in Humans

Organism	Disease
	Bacteria
Campylobacter jejuni	Bloody diarrhea, abdominal pain
Bacillus anthracis	Skin disease
Brucella abortus	Gastrointestinal disease, anorexia
Escherichia coli	Gastrointestinal disease
Leptospira spp.	Kidney infection
Listeria monocytogenes	Meningitis in adults, infection in infants
Mycobacterium tuberculosis	Tuberculosis
Mycobacterium paratuberculosis	Johnes disease
Salmonella spp.	Salmonellosis
Yersinia enterocolitica	Gastrointestinal infection
Streptococcus pyrogenes	Infections
Staphylococcus aureus	Infections
Chlamydia spp.	Trachoma, conjunctivitis, urogenital tract infections, pneumonia, bronchitis, infections
Clostridium spp.	Infections
	Viruses
Adenoviruses	Eye and respiratory infections
Avian enteroviruses	Respiratory infections
Avian reoviruses	Infectious bronchitis
Bovine parovirus	Respiratory infection
Bovine rhinovirus	Foot and mouth disease
Enterovirus	Respiratory infection
Reovirus	Respiratory infection
Rhinovirus	Parainfluenza
Rotaviruses	Gastrointestinal infections
	Parasites
Protozoa	
Balatidium coli	Blantidias
Cryptosporidium parvum	Cryptosporidiasis
Eimeria spp.	Coccidiosis
Giardia lamblia	Giardiasis
Toxoplasma spp.	Toxoplasmosis
	Helminth Worms
Ascaris lumbricoides	Ascariasis
Trichuristrichiura	Whip worm infection
	Prions
Mad cow disease	Neurological degeneration

Source: Epstein, E. *The Science of Composting*, CRC Press, Boca Raton, FL, 1997; Smith, J. E. and Epstein, E., in *WEFTEC '99—Conference Workshop: Beneficial Use of Animal Residuals*, New Orleans, LA, Water Environment Federation, Arlington, VA, 1999; Bowman, D. D. *Manure Pathogens: Manure Management, Regulations, and Water Quality Protection*, WEF Press, Alexandria, VA, 2009.

TABLE 12.4
Estimate of Indicator Bacteria in Various Components of Solid Waste

	Total Coliforms %		Fecal Coliforms %		Fecal Streptococci %	
Waste Category	**1974**	**1975**	**1974**	**1975**	**1974**	**1975**
Paper	66.4	33.2	29.5	13.5	14.6	51.4
Garden waste	18.8	18.3	6.0	7.4	54.0	14.6
Metal	0.5	4.1	10.5	5.3	4.2	7.3
Glass	0.4	0.1	4.3	0.5	0.9	0.1
Food waste	2.1	19.6	2.9	8.4	2.5	13.4
Plastic, rubber, leather	0.6	7.4	1.8	10.6	0.9	5.6
Textiles	6.0	3.1	1.6	4.0	1.4	0.5
Fines	4.8	8.2	42.6	4.1	16.4	2.6
Ash, rock, dirt	0.3	1.5	0.2	10.2	3.6	2.3
Diapers	0.06	4.5	0.6	35.9	1.3	2.2
Wood	0.01	0.0	0.0	0.1	0.2	0.0

Source: Pahren, H. R. *CRC Crit. Rev. Environ. Control*, 17(3), 187–228, 1987. With permission.

TABLE 12.5
Microbial Concentrations in Food Waste, Yard Waste, and Wood Waste

Organism	**Food Waste**	**Yard Waste**	**Yard Waste and Waste Paper**	**Wood Waste**
		Indicator Organisms		
Total coliforms	5.00×10^6	8.0×10^5	5.00×10^5	1.30×10^6
Fecal coliforms	2.00×10^4	8.00×10^5	5.00×10^5	1.30×10^6
Escherichia coli	3.50×10^3	8.00×10^5	3.00×10^5	1.30×10^6
Fecal streptococci	8.00×10^6	1.60×10^6	1.60×10^6	1.60×10^6
Enterococcus	1.30×10^5	2.30×10^5	1.30×10^5	3.00×10^5
		Pathogens		
Salmonella spp.	<0.002	0.002	0.36	
Staphylococcus	32.2	0.8	4.4	3.8
Listeria spp.	<0.02	0.02	<0.02	<0.02
Parasites	Protozoa	Negative	Negative	Negative

Source: E&A Environmental Consultants, Inc., *Food Waste Collection and Composting Demonstration Project for City of Seattle Solid Waste Utility*, Seattle, WA, 1994.

(Morris and Feeley, 1976), and *Listeria monocytogenes* (van Renterghem et al., 1991), and viruses, *Giardia lamblia, Campylobacter jejuni, Cryptosporidium parvum,* and *Cyclospora* (Millner, 1998). Hess et al. (2004) reported that *E. coli* O157:H7 survived in raw manure outdoors for twenty-one months.

There is no evidence or scientific literature indicating human health outbreaks from land application of biosolids or compost. This is most likely because of the effectiveness of the USEPA regulation 40CFR503 (USEPA, 1994).

In 2007, a major contamination of spinach with *E. coli* O157:H7 resulted in deaths and illnesses. It was determined that runoff and irrigation of water contaminated from raw manure were the cause. Food waste and yard waste will generally contain human and animal pathogens. Raw meats, contaminated food, and spoiled fruits and vegetables can harbor pathogens. Yard wastes often contain pathogens as a result of domesticated pet excrement.

Although there are numerous human pathogenic organisms found in manures (Table 12.3), relatively few organisms have been linked to the use of untreated manures. Staley and Haines (2009) indicate that the application of untreated manure to land may result in significant pathogen contamination of surface and ground water.

TYPES OF PATHOGENS

This section describes the most common and predominant foodborne and waterborne diseases in the United States (CDC, 2005). The predominant pathogens are *Salmonella* spp., *Escherichia coli* O157:H7, *Cryptosporidium parvum, Listeria monocytogenes, Mycobacteriaum paratuberculosis,* and *Giardia* spp.

SALMONELLA SPP.

The genus *Salmonella* is a member of the family Enterobacteriaceae. Salmonella are complex organisms that produce a variety of virulent factors, including endotoxin, cytotoxins, and enterotoxins (Joklik et al., 1992). The toxins produced by *S. typhimurium* are verotoxin and enterotoxin. Clinical manifestations are gastroenteritis, septicemia, or an enteric fever such as typhoid fever (Joklik et al., 1992; Ohl and Miller, 2001; Debbie, 2009). The organism is fairly sturdy and can survive in the environment for extended periods of time (Guan and Holley, 2003).

Salmonella spp., especially *S. typhi* and *S. typhimurium,* infections have increased over the past thirty years (Tauxe, 1997). It is estimated that over 2 million cases occur in the United States each year. Much of the increase is the result of antimicrobial resistance among *Salmonella* spp. serotypes (Pell, 1997). Pell (1997) indicates that one of the consequences of antimicrobial resistance is that *Salmonella* spp. become a larger proportion of the microbial population because competing organisms are unable to grow, thus increasing the risk of infection. The human infective dose of *S. typhimurium* is 10^7 to 10^9 cells per gram. Major foodborne outbreaks have been related to seafood, beef, poultry, eggs, milk and dairy products, sauces and salad dressing, homemade ice cream, and pork (FDA, 2009).

Salmonella spp. have been identified in cattle feces (Jones, 1980), poultry manure (Henzler et al., 1998), pig slurry (Strauch, 1991), and horses (Murray, 1998). Jones (1980) reported that healthy cattle might excrete as many as 10^7 salmonellae per gram of feces. Henzler et al. (1998) found *S. enteritidis* in manure from poultry flocks, and that flocks with high levels of manure contamination were ten times more likely to produce contaminated eggs. Diarrheic horses present the greatest risk of *Salmonella* spp. contamination to the environment (Murray, 1998).

The main source of infection of *Salmonella enterica* has been associated with ingestion of food such as alfalfa sprouts, unpasteurized orange juice, and contaminated tomatoes (Debbie, 2009).

Escherichia coli O157:H7

E. coli is in the family of Enteribacteriaceae. They are gram-negative, rod-shaped bacteria, and can be nonmotile or motile by means of flagella (USDA, 1994).

E. coli are a normal inhabitant of the lower intestines of warm-blooded animals. The verotoxin-producing *E. coli* O157:H7 was first identified as a human pathogen in 1982. The verotoxins produced by this organism can result in three different sets of symptoms: hemorrhagic colitis (diarrhea that becomes profuse and bloody), hemolytic uremic syndrome (bloody diarrhea followed by renal failure), and thrombocytopenic purpura (symptoms similar to those of hemolytic uremic syndrome with central nervous system involvement). Death often occurs in patients with hemolytic uremic syndrome and thrombocytopenic purpura (Pell, 1997). The human infective dose is believed to be $\leq 10^2$ cells/gram (USDA, 1994).

Escherichia coli O157:H7 has been documented in dairy cattle and implicated as a principal reservoir in young animals. However, several studies have shown that relatively few fecal samples contained the organism (Zhao et al., 1995; Garber et al., 1995, 1999). In general, between 1 and 5% of the animals harbored the organism. In their most recent study, Garber et al. (1999) reported that of the 4,361 dairy cows on 91 dairy operations, only 52 (1.2%) of the fecal samples from 22 (24.2%) of the operations contained O157:H7.

Kudva et al. (1998) studied the survival of O157:H7 in manure. He found that the organism survived for at least one hundred days in bovine frozen manure. At temperatures from 4 to 70°C the survival ranged from forty days to twenty-four hours.

In the summer of 2006, *E. coli* O157:H7 infected over 120 persons from the consumption of bagged spinach. During the past decade, there have been twenty reported cases of infection by this organism from the consumption of lettuce, spinach, and other leafy vegetables. It is estimated by the Centers for Disease Control that as many as twenty thousand cases occur annually. Several causes for contamination have been suggested. These include contaminated water by animal feces, the use of improperly composted manure (composting of manure is not regulated), and direct application of manure. There have been several small and large waterborne outbreaks because of water contamination by animal manures (Debbie, 2009).

Other Pathogens

Recently several other pathogens and pathogenic substances have emerged as a concern for human health. These organisms and diseases are often referred to as emerging diseases, even though many of them are not new. Examples include:

- *Helicobacter pylori*
- *Legionella* spp.
- *E. coli* O157:H7
- *Listeria monocytogenes*
- Human immunodeficiency virus (HIV)
- *Cryptosporidium parvum*
- *Campylobacter* spp.

Helicobacter pylori is a gram-negative, spiral microaerophilic bacterium. It is associated with gastric and duodenal ulcers and is considered to be a risk for gastric cancer. It causes abdominal pain, anorexia, nausea, and vomiting. It possibly is the most prevalent bacterial infection in man and the most chronic infection of humans. Up to 90% of some populations are infected. Contamination is believed to be through food and water (CDC, 2005).

Legionella spp. is ubiquitous in the environment. It is found naturally in the environment, usually in water. *Legionella* spp. can cause respiratory infection. Symptoms are high fever, chills, and cough. It can cause death. It has been associated with gardening, compost, and the use of potting soil in Australia and Japan. There have been three cases in the United States associated with potting soil. It is easily destroyed by high temperatures achieved through proper composting.

Listeria monocytogenes is a gram-negative bacterium. It is found in soil, silage, and on plants. It causes foodborne diseases and has been associated with raw milk, cheeses, ice cream, raw vegetables, raw meat, poultry, and raw and smoked fish. It can cause septicemia, meningitis, encephalitis, and intrauterine or cervical infections. It does not affect healthy people (McLauchin et al., 2004).

HIV belongs to a group of retroviruses that infect and destroy helper T cells of the immune system. It is the cause of AIDS (acquired immunodeficiency syndrome), which infected over 25 million persons.

Cryptosporidium parvum is a protozoan parasite. The acute disease is intestinal, tracheal, or pulmonary cryptosporidiosis. It causes diarrhea, vomiting, and weight loss. Farm animals are the major contributors. Infected calves may shed 1 million oocysts per day during the first twelve days and can result in soil and water resource contamination (Sischo, 2000; Walker and Stedinger, 1999). Atwill et al. (2003) indicated that adult beef cattle may shed between three thousand and nine thousand oocysts per cow per day.

Campylobacter jejuni is a gram-negative microaerophilic organism. It is probably the most common enteric bacterial infection causing bacterial diarrheal illness in the United States. Other symptoms are fever, abdominal pain, nausea, and muscular pain. Incidences exceed *Salmonella* and *Shigella* infections combined. The number of cases is estimated to exceed 2 million to 4 million per year.

Mad cow disease or bovine spongiform encephalophy (BSE) is a neurological, transmissible degenerative disease believed to be caused by a proteinaceous infectious particle termed prion (Epstein and Beecher, 2005). Several animal and human neurological diseases are attributed to prions. These include scrapies in sheep, mad cow disease (BSE) in cattle, chronic wasting disease (CWD) in deer and elk, kuru in humans (primarily as a result of cannibalism through the consumption of infectious tissues), Creutzfeld-Jacob (CDJ) disease in humans, fatal familial insomnia (FFI), and Gerstmann-Straussler-Scheinker (GSS). BSE is characterized by rapidly progressive dementia. Patients experience impaired memory, impaired vision, and muscular coordination. They may also experience insomnia, depression, or unusual sensations. As the illness progresses, mental impairment becomes severe and blindness may occur. Eventually, the patient may lose the ability to move and speak. Other infections, such as pneumonia, can set in and lead to death (Epstein and Beecher, 2005).

PUBLIC AND WORKER CONCERNS AND ISSUES

The public residing adjacent to composting facilities may become concerned with exposure to pathogens. This is especially true when the facility composts sewage sludge or biosolids. Generally, the public is less concerned with pathogens associated with manure, yard waste, or other feedstocks, even though they may contain high levels of pathogens. This is because of the knowledge that human wastes contain pathogens and lack of realization that other wastes may contain pathogens. Earlier in this chapter, it was clearly shown that raw manure contains high levels of human pathogens. The potential exposure is greatest to workers handling wastes prior to composting.

Risk assessment models are currently the best methods to assess the risk of infection from exposure to aerosols (Hass et al., 1999; Brooks et al., 2004). In a risk assessment of land application of biosolids, Brooks et al. (2004) showed that the risk to the public is minimal. The health risk to workers is greater, and proper hygiene practices are important. There are no risk assessment data for composting operations.

There are four major aspects in the design and operation of composting facilities with regard to pathogen destruction:

- Control of pathogen emissions and dispersions
- Temperature control for effective pathogen destruction
- Prevention of regrowth
- Prevention of recontamination

Control of Pathogens, Their Emission, and Dispersion during Composting

Generally, the emissions and dispersion of primary pathogens during composting operations do not occur to any extent. Dispersion distance is short and does not extend beyond the facility. Therefore, the potential for the public being infected by residing near a facility is very small. The problem is of importance with regard to bioaerosols, and this subject will be discussed in Chapter 13.

The control of pathogens during composting is of paramount importance. This control is achieved through maintaining proper time-temperature. As indicated earlier, USEPA regulates the time-temperature requirements for sewage sludge and biosolids. Many states apply these criteria to other feedstocks. Below is a summary of the USEPA regulations (USEPA, 2003).

1. Process requirements
    - Using either the within-vessel composting method or the static aerated pile composting method, the temperature of sewage sludge is maintained at 55°C (131°F) or higher for three consecutive days.
    - Using the windrow composting method, the temperature of the sewage sludge is maintained at 55°C (131°F) for fifteen consecutive days or longer. During the period when the compost is maintained at 55°C (131°F) or higher, there shall be a minimum of five turnings of the windrow.
2. Pathogen testing requirements
    - Either the density of fecal coliforms in the sewage sludge must be less than 1,000 MPN per gram of total solids (dry weight basis), or the *Salmonella* sp. bacteria in sewage sludge must be less than 3 MPN per 4 g of total solids (dry weight basis) at the time the sewage sludge is used or disposed, or at the time the sewage sludge is prepared for sale or given away in a bag or other container for land application.
    - The density of enteric viruses in the sewage sludge after pathogen treatment must be less than 1 PFU per 4 g of total solids (dry weight basis).
    - The density of viable helminth ova in the sewage sludge after pathogen treatment must be less than 1 per 4 g of total solids (dry weight basis).

Only the analysis for fecal coliforms or *Salmonella* spp. is needed if proper time-temperatures are monitored and recorded. Otherwise, additional testing for enteric viruses and helminth ova must be done. These are difficult and expensive. The effectiveness of time-temperature has been demonstrated by several studies.

Brooks et al. (2004) state that the potential for adverse health affects from pathogens in aerosols depends on their fate and transport. During composting pathogens are destroyed. This can occur in as little as three days at 55°C (131°F), depending on the technology. The potential for transport also is a function of the technology. A system that agitates the compost will result in greater dispersion and transport. Brooks et al. (2004) indicate there are numerous environmental factors, such as relative humidity, wind velocity, and method of aerosol generation, that affect transport. Relative humidity, temperature, ultraviolet radiation, oxygen concentration, and method of aerosol generation affect their fate. Bacterial inactivation results from dehydration, desiccation, and elevated temperatures. Brooks et al. (2004) state that despite the generation of aerosols, the microbes within the aerosols either are inactivated or fail to be transported over any significant distance.

Consideration needs to be given for worker protection. Bünger et al. (2000) reported that workers at composting sites were exposed to total bacteria, as well as other bioaerosols. Good hygiene is very important. Workers injured or having open wounds need to be especially vigilant. Workers need to be encouraged to seek basic and simple first aid when injured, regardless of the size of the injury. Immunosuppressed individuals should not be employed in operations where they can be exposed to material containing pathogens. They should be informed of the dangers and make sure their physician approves the area of employment.

Effect of Composting on Pathogen Destruction

The time-temperature criteria are most important for effective pathogen destruction during composting. One of the best examples of this relationship is the pasteurization of milk. Heating raw milk to 60 to 63°C (140 to 145°F) for twenty to thirty minutes results in pasteurization and destruction of pathogens. In addition to the effect of temperature during composting, ammonia is released. Ammonia is also a disinfecting agent. Furthermore, as will be shown, the presence of indigenous bacteria also reduces the level of pathogens.

Stern (1974) in Table 12.6 provided data on the time-temperature requirement for the destruction of numerous pathogens. Another study showing the destruction of poliovirus in 40% compost material, as related to time and temperature, is shown in Table 12.7. It is evident that temperature is very effective in pathogen destruction, and that the time to destroy various pathogens is affected by the length of exposure to temperature. At 35°C (95°F) poliovirus survival was much longer than at 47°C

TABLE 12.6
Time-Temperature Required for the Destruction of Several Pathogens in Biosolids

	Exposure Time in Minutes for Destruction of Pathogens at Various Temperatures				
Organism	**50°C**	**55°C**	**60°C**	**65°C**	**70°C**
Entamoeba hystolytic cysts	5				
Ascaris lumbricoides eggs	60	7			
Brucella abortus		60		3	
Corynebacterium diphtheria		45			4
Salmonella typhi			30		4
Escherichia typhi			60		5
Microccus pyogenes var. *aureus*					20
Mycobacterium tuberculosis					20
Shigella spp.	60				
Mycobacterium diphtheria	45				
Necator americanus	50				
Taenia saginata					5
Viruses					25

Source: Stern, G. 1974. Pasteurization of liquid digested sludge. In Proceedings of the National Conference on Composting Municipal Sludge Management. Silver Spring, MD: Information Transfer, Inc.

TABLE 12.7
Inactivation of Poliovirus in Composted Biosolids at 60% Moisture

Treatment	Percentage Recovery of Plaque-Forming Units (PFUs)
35°C, 200 minutes	30
39°C, 20 minutes	7.2
43°C, 20 minutes	0.087
47°C, 5 minutes	0.003

(117°F). During studies on effectiveness of temperature to destroy pathogens in compost, Ward and Brandon (1977) investigated the heat inactivation of coliform bacteria, fecal streptococcus bacteria, and *Salmonella enteritidis.* The data are shown in Figures 12.2 to 12.4. As a result of the hardiness survival of fecal coliform bacteria and *Salmonella* spp., the USEPA and other scientists felt that the destruction of these organisms would be good indicators showing that other pathogens in the compost would be destroyed. The USEPA uses a time-temperature formula to assess the time required at a given temperature for pathogen inactivation.

This equation is

$$D = 131{,}700{,}000/10^{0.1400t}$$

where:

D = time in days
t = temperature in degrees Celsius

As indicated earlier, product disinfection is a major goal of a good composting operation. The public obtaining the product must be comfortable that the product is safe. The major method of disinfection is maintaining proper time-temperature. In my previous book, *The Science of Composting* (Epstein, 1997), data were shown on the effectiveness of the composting operation on inactivation and destruction of pathogens. Walke (1975) monitored *Escherichia coli*, *Salmonella eidleberg*, and *Candia albicans* during windrow composting of bark-biosolids mixtures. The initial compost contained these organisms at a level of 10^6 microbes per dry gram. After twenty-four hours, the levels were 11, 130, and 620 microbes per dry gram of solids for *E. coli, Salmonella* spp., and *Candia albicans*, respectively. No organisms were detected after thirty-six hours. Since then, there have been relatively little new data available.

In Figures 12.2 to 12.4, the rate of inactivation of fecal coliform, fecal streptococcus, and *Salmonella enteritidis* is shown as temperature increases.

Senne et al. (1994) reported that the HPA1 virus and adenovirus 76 were inactivated during ten to twenty days of composting of poultry carcasses. Lung et al.

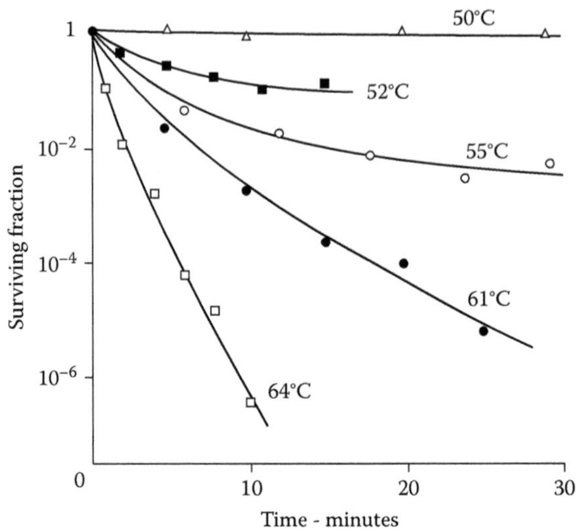

FIGURE 12.2 Heat inactivation of coliform bacteria in composted biosolids. (Data from Ward, R. L. and Brandon, J. R. in *National Conference on Composting Municipal Residues and Sludges*, Information Transfer, Rockville, MD, 1977, 122–34).

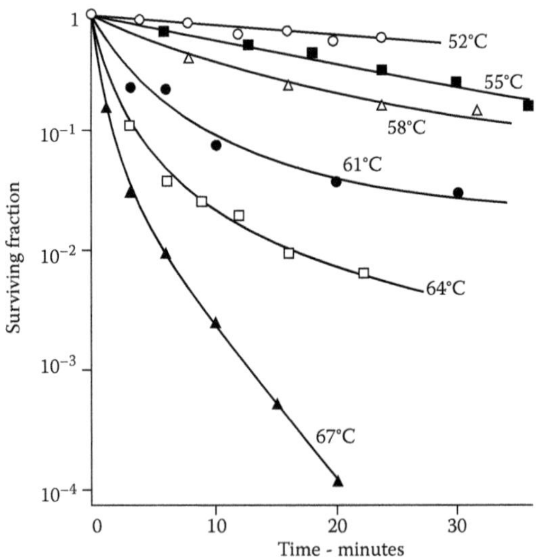

FIGURE 12.3 Heat inactivation of fecal streptococcus in composted biosolids. (Data from Ward, R. L. and Brandon, J. R. in *National Conference on Composting Municipal Residues and Sludges*, Information Transfer, Rockville, MD, 1977, 122–34).

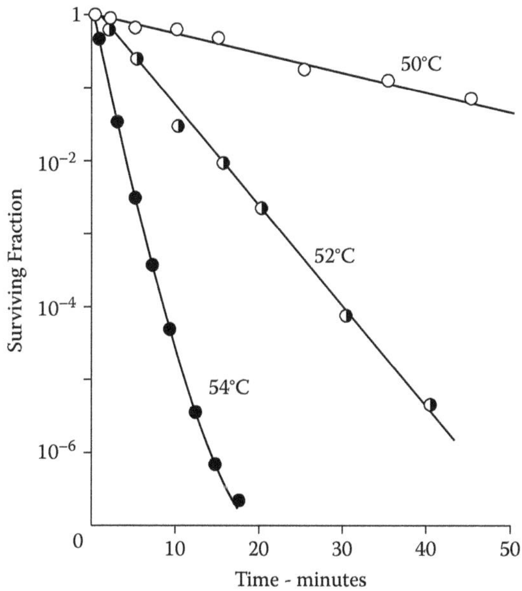

FIGURE 12.4 Heat inactivation of *Salmonella enteritidis* serotype Montevideo in composted biosolids. (Data from Ward, R. L. and Brandon, J. R. in *National Conference on Composting Municipal Residues and Sludges*, Information Transfer, Rockville, MD, 1977, 122–34).

(2001) evaluated the effect of composting on *E. coli* O157:H7 and *Salmonella enteritidis* during cow manure composting. *E. coli* O157:H7 was not detected after seventy-two hours of composting at 45°C (113°F), and *Salmonella enteritidis* was not detected after forty-eight hours. Lung et al. (2001), in a bench-scale composting study, reported that *E. coli* O157:H7 was destroyed at 47°C (117°F) in seventy-two hours. Hess et al. (2004) found that *E. coli* O157:H7 from infected cows was not detected in twenty-four hours at composting temperatures of 55°C (131°) and 65°C (149°F). Additional substantiation on the thermal inactivation of *E. coli* O157:H7 was provided by Jiang et al. (2003) and Morgan and Doyle (2003).

Grewal et al. (2006) did not detect *E. coli*, *Salmonella*, and *Listeria* found in manure after three days of composting at 55°C (131°F). They recommended thermophilic composting of manure destined for vegetable production, residential gardening, or application to rapidly draining fields. In a subsequent study, Grewal et al. (2007) reported that thermophilic composting destroyed *Listeria* and *Salmonella* in swine manure.

Operators must maintain records. When composting sewage sludge and biosolids, not only must time-temperature records be documented, but either *Salmonella* or fecal coliforms must be determined. In the event that the time-temperature records are not recorded or are deficient, the product must be examined not only for *Salmonella* and fecal coliform bacteria, but also for viruses and helminth ova. The latter two are very expensive to test for, and few laboratories are capable of their determination. For other feedstocks, state regulations must be adhered.

Bacterial Regrowth and Recontamination

One issue with pathogenic bacteria is the potential for regrowth (Epstein, 1997). Viruses and helminth ova do not regrow. Obligate parasites inactivated in one phase of the composting process cannot regrow (Haug, 1993). Avoiding the potential for regrowth of bacteria requires the removal of their food source. Indigent organisms are much hardier than most pathogenic organisms. The best method to minimize the potential for regrowth is to make sure the compost is stabilized. As Haug (1993) indicates, the healthy and diverse microbial population of a well-stabilized compost should suppress the regrowth of pathogens. During the time of curing, stabilization occurs, which lowers the nutrient content of the media necessary for the regrowth of pathogens.

Achieving stabilization of the final product is important to control regrowth. Epstein (1997) discussed the aspect of stabilization and its measurement. I indicated that the term *stabilization* is often used synonymously with *maturity*. Stabilization is a function of the process and denotes the extent the process is carried out and to what extent decomposition occurs. Maturity is a chemical index and indicates the potential impact of the product on plant germination and growth. Stabilization is achieved through proper curing or residence time. In the process, there is a diminishing of the food source, but also the reestablishment of the indigenous microflora (Haug, 1993). Both these conditions are not conducive to the regrowth of pathogens.

Recontamination of the cured product must be avoided. In large operations where separate equipment is used for the various activities, there is little potential that equipment is used for handling the feedstock, as well as the final product. Small operations do not have that luxury. Therefore, they need to disinfect the equipment between operations. For example, the use of a power washer containing bleach can be effectively used for disinfecting the bucket and tires of a front-end loader.

Worker Exposure and Prevention

As was indicated above, the public will not be impacted by the potential dispersion of pathogens during the handling and composting of feedstocks. There is, however, a potential for workers to be infected primarily during the handling of feedstocks containing pathogens. The best prevention is good hygiene. Workers should avoid eating or smoking unless they wash their hands. Cuts and any skin openings should be treated. Inhalable dust should be avoided and masks should be worn where dusty operations occur. Prevention is the key for exposure to pathogens.

CONCLUSION

Most feedstocks contain pathogens. Although the USEPA mandates pathogen and vector attraction reduction for sewage sludge and biosolids, the same regulations should be applied to manures. Land application of untreated or raw manures should be avoided if vegetables are grown on home gardens or lawns.

Time-temperature control during the composting process is imperative to produce a disinfected product. Good data would provide regulators and the public with confidence that the composting process is carried out properly.

The composting process will not affect the public residing near composting facilities as a result of pathogens in the feedstock. Dispersion is minimal and rarely extends beyond the property boundary of a facility.

Workers are more vulnerable and need to exercise precautions by practicing good hygienic practices.

REFERENCES

Akin, E. W. 1983. *Infective dose of waterborne pathogens*. Cincinnati, OH: Health Effects Research Laboratory.

Atwill, E. R., Hoar, B., das Gracas Cabral Pereira, M., Tate, K. W., Rulofson, F., Nader, G., and Das, G. C. P. M. 2003. Improved quantitative estimates of low environmental loading and sporadic periparturient shedding of *Cryptosporidium parvum* in adult beef cattle. *Appl. Environ. Microbiol.* 69(8):4604–10.

Bicudo, J. R., and Goyal, S. M. 2003. Pathogens and manure management systems: A review. *Environ. Technol.* 24(1):115–30.

Bitten, G. 1994. *Pathogens and parasites in domestic wastewater*. In Wastewater microbiology, ed. G. Bitten. New York: Wiley & Sons.

Bowman, D. D., ed. 2009. *Manure pathogens: Manure management, regulations, and water quality protection*. Alexandria, VA: WEF Press.

Brooks, G. F., Butel, J. S., and Ornston, L. N. 1991. *Jawetz, Melnick, and Adelberg's medical microbiology*. 19th ed. Norwalk, CT: Appleton and Lange.

Brooks, J. P., Gerba, C. P., and Pepper, I. L. 2004. Biological aerosol emission, fate, and transport from municipal and animal wastes. *J. Residuals Sci. Technol.* 1(1):15–28.

Bryan, F. L. 1977. Disease transmitted by foods contaminated by wastewater. *J. Food Protect.* 40:45–52.

Bünger, J., Antlauf-Lammers, M., et al. 2000. Health complaints and immunological markers of exposure to bioaerosols among biowaste collectors and compost workers. *Occup. Environ. Med.* 57:458–64.

Casemore, D. P. 1991. The epidemiology of human cryptosporidiosis and the water route of infection. *Water Sci. Technol.* 24(2):157–64.

CDC. 2005. Foodborne illnesses. Washington, DC: Centers for Disease Control, U.S. Department of Health and Human Services. Available at http://www.cdc.gov/ncidod/dbmd/diseaseinfo/foodborneinfections_g.htm.

Cliver, D. O. 1980. Infection with minimal quantities of pathogens from wastewater aerosols. In *Wastewater, aerosols, and disease. Proceedings of a symposium*, ed. H. Pharen and W. Jakubowski. Cincinnati, OH: USEPA.

Cole, D. J., Hill, V. R., Humenik, F. J., and Sobsey, M. D. 1999. Health safety and environmental concerns of farm animal waste. *Occup. Med.* 14:423–48.

Craun, M. F., Graun, F. G., Claderon, R. L., and Beach, M. J. 2006. Waterborne outbreaks reported in the United States. *J. Water Health* 4(Suppl. 2).

Debbie, D. P. 2009. Bacterial pathogens in animal manure. In *Manure pathogens: Manure management, regulations, and water quality protection*, ed. D. D. Bowman. Alexandria, VA: WEF Press and McGraw Hill.

Derbyshire, J. B. 1976. Fate of animal viruses in effluent from liquid farm wastes. *J. Milk Food Technol.* 39(3):214–17.

E&A Environmental Consultants, Inc. 1994. *Food waste collection and composting demonstration project for City of Seattle Solid Waste Utility*. Seattle, WA.

Epstein, E. 1997. *The science of composting*. Boca Raton, FL: CRC Press.

Epstein, E., and Beecher, N. 2005. Mad cow disease, Creuzfeldt-Jakob disease, and other TSEs and biosolids. *J. Residuals Sci. Technol.* 2(3):181–87.

Epstein, E., Servo, S., Ettlin, L., Cutler, A., and Sullivan, D. 2005. Composting of commercial food waste: Pathogen and related environmental issues. *J. Residuals Sci. Technol.* 2(2):79–89.

FDA. 2009. *Bad Bug Book, Foodborne Pathogenic Microorganisms and Natural Toxins Handbook*, FDA, Washington, DC.

FDA. 2009. Washington, DC: Centers for Disease Control, U.S. Department. of Health and Human Services. Available at http://www.fda.gov/food/foodsafety/foodborne illnesses.

Garber, L., Wells, S., Schroeder-Tucker, L., and Ferris, K. 1999. Factors associated with fecal shedding of verotoxin-producing *Escherichia coli* O157:H7 on dairy farms. *J. Food Protect.* 62(4):307–12.

Gerba, C. P. 1996. Microbial pathogens in municipal solid waste. In *Microbiology of solid waste*, ed. A. C. Palmisano and M. A. Barlaz. Boca Raton, FL: CRC Press.

Gerba, C. P., Huber, M. S., Naranjo, J., Rose, J. B., and Bradford, S. 1995. Occurrence of enteric pathogens in composted domestic solid waste containing disposable diapers. *Waste Manage. Res.* 13:315–24.

Grewal, S. K., Rejeev, S., Sreevatsan, S., and Michel Jr., F. C. 2006. Persistence of *Mycobacterium avium* subsp. *paratuberculosis* and other zoonotic pathogens during simulated composting, manure packing, and liquid storage of dairy manure. *Appl. Environ. Microbiol.* 72:565–74.

Grewal, S., Sreevatsan, S., and Michel Jr., F. C. 2007. *Listeria* and *Salmonella* during swine manure management. *Compost Sci. Util.* 15:53–62.

Guan, T. Y., and Holley, R. A. 2003. Pathogen survival in swine manure environments and transmission of human enteric illnesses—A review. *J. Environ. Qual.* 32:383–92.

Hass, C. N., Rose, J. B., and Gerba, C. P., ed. 1999. *Quantified microbial risk assessment*. New York: John Wiley & Sons.

Haug, R. T. 1993. *The practical handbook of compost engineering*. Boca Raton, FL: Lewis Publishers.

Hess, T. F., Grdzelishvili, I., Sheng, H., and Hovede, C. J. 2004. Heat inactivation of *E. coli* during manure composting. *Compost Sci. Util.* 12(4):314–22.

Henzler, D. J., Kradel, D. C., and Sischo, W. H. 1998. Management and environmental risk factors for *Salmonella enteritidis* contamination of eggs. *Am. J. Vet. Res.* 59(7):824–29.

Hornick, R. B., Greisman, S. E., Woodward, T. E., Dupont, H. L., Dawkins, A. T., and Snyder, M. J. 1970. Typhoid fever: Pathogenesis and immunological control. *New Engl. J. Med.* 283:686–91.

Huber, M. S., Gerba, C. P., et al. 1994. Study of persistence of enteric viruses in landfilled disposable diapers. *Environ. Sci. Technol.* 28:176–77.

Jiang, X., Morgan, J. X., and Doyle, M. P. 2003. Thermal inactivation of *Escherichia coli* O157:H7 in cow manure compost. *J. Food Protect.* 66:1771–77.

Joklik, W. K., Willett, H. P., Amos, D. B., and Wilfert, C. M. 1992. *Zinsser microbiology*. 20th ed. Norwalk, CT: Appleton & Lange.

Jones, D. L. 1999. Potential health risks associated with the persistence of *Escherichia coli* O157:H7 in agricultural environments. *Soil Use Manage.* 15:76–83.

Jones, P. W. 1980. Health hazards associated with the handling of animal wastes. *Vet. Rec.* 106:4–7.

Keswick, B. H. 1984. Sources of groundwater pollution. In *Groundwater pollution microbiology*, ed. G. Bitton and G. Gerba. New York: John Wiley & Sons.

Kowal, N. E. 1982. *Health effects of land treatment: Microbiological*. Cincinnati, OH: USEPA Health Effect Research Laboratory.

Kowal, N. E. 1985. *Health effects of land application of municipal sludge*. EPA/600/1-85/015. Cincinnati, OH: U.S. Environmental Protection Agency.

Kudva, I. T., Blanch, K., and Hovde, C. J. 1998. Analysis of *Escherichia coli* O157:H7 survival in ovine or bovine manure or manure slurry. *Appl. Environ. Microbiol.* 64(9):3166–74.

Levine, M. M., Dupont, H. L., et al. 1973. Pathogenesis of *Shigella dysenteriae* (Shiga) dysentaery. *J. Infect. Dis.* 127(3):261–70.

Lung, A. J., Lim, C. M., Kim, J. M., Marshall, M. R., Nordstedt, R., Thompson, N. P., and Wei, C. I. 2001. Destruction of *Escherichia coli* O157:H7 and *Salmonella enteritidis* in cow manure composting. *J. Food Protect.* 64:1309–14.

Mackenzie, W. R., Hoxie, N. J., Proctor, M. E., Gradus, M. S., Blair, K. R., Peterson, D. E., Kazmierczak, J. J., Addiss, D. G., Fox, K. R., Rose, J. B., and Davis, J. P. 1994. A massive outbreak in Milwaukee of *Cryptosporidium* infection transmitted through the public water supply. *New Engl. J. Med.* 331:161–67.

McLauchin, J., Mitchell, R. T., Smerdon, W. J., and Jewell, W. J. 2004. *Listeria monocytogenes* and listeriosis: A review of hazard characteristics for use in microbiological risk assessment of foods. *Int. J. Food Microbiol.* 92:15–33.

Millner, P. D. 1998. Microbial effects on environmental health and product quality aspects of recovery and co-utilization of bio-mineral products. In *Beneficial Co-utilization of agricultural, municipal, and industrial by-products*, ed. S. Brown, J. S. Angle, and L. Jacobs, 247–58. Amsterdam: Kluwer Academic Publishers.

Millner, P. D., and Karnes, J. 2005. Animal manure: Bacterial pathogens and disinfection technologies. Contemporary perspectives on infectious disease agents. In *Sewage sludge and manure*, ed. J. E. J. Smith, P. D. Millner, W. Jakubowski, N. Goldstein, and R. Rynk. Emmaus, PA: JG Press, Inc.

Morgan, J. X., and Doyle, M. P. 2003. Thermal inactivation of *Escherichia coli* O157:H7 in cow manure compost. *J. Food Protect.* 10:1771–77.

Morris, G. K., and Feeley, J. C. 1976. *Yersinia enterocolitica*: A review of its role in food hygiene. *Bull. World Health Org.* 54:57–85.

Murray, M. J. 1998. How should clients manage horses that have diarrhea and cultured positive for *Salmonella* to minimize exposure to other horses? *Compendium Continuing Educ. Practicing Vet.* 20(12):1352.

Ohl, M. E., and Miller, S. I. 2001. *Salmonella*: A model for bacterial pathogenesis. *Ann. Rev. Med.* 52:259–74.

Okhuysen, P. C., Chappell, C. L., Crabb, J. H., Sterling, C. R., and Dupont, H. L. 1999. Virulence of three distinct *Cryptosporidium parvum* isolates for healthy adults. *J. Infect. Dis.* 180:1275–81.

Pahren, H. R. 1987. Microorganisms in municipal solid waste and public health implications. *CRC Crit. Rev. Environ. Control* 17(3):187–228.

Payment, P. 1989. Presence of human and animal viruses in surface and ground waters. *Water Sci. Technol.* 21(3):283–85.

Pell, A. N. 1997. Manure and microbes; public and animal health problem. *J. Dairy Sci.* 89:2673–81.

Robinson, D. A. 1981. Infective dose of *Campylobacter jejuni* in milk. *Br. Med. J.* 282:1584.

Senne, D. A., Parnigrahy, B., and Morgan, R. L. 1994. Effect of composting poultry carcasses on survival of exotic avian viruses: Highly pathogenic avian influenza (HPAI) virus and adenovirus of egg drop syndrome-76. *Avian Dis.* 38:733–37.

Sischo, W. M., Atwill, E. R., Lanyon, L. E., and George, J. 2000. *Cryptosporidium* on dairy farms and the role these farms may have in contaminating surface water supplies in the northeastern United States. *Prev. Med.* 43(4):253–67.

Smith, J. E. J., and Epstein, E. 1999. Issues and concerns from the viewpoint of the USEPA's Office of Research and Development. In *WEFTEC '99—Conference Workshop: Beneficial Use of Animal Residuals*, New Orleans, LA. Arlington, VA: Water Environment Federation.

Smith, J. E., Millner, P. D., Jakubowski, W., Goldstein, N., and Rynk, R., ed. 2005. *Contemporary perspectives on infectious disease agents in sewage sludge and manure*. Emmaus, PA: JG Press, Inc.

Staley, L., and Haines, J. 2009. Environmental pollution and concentrated animal feeding operations. In *Manure pathogens: Manure management, regulations, and water quality protection*, ed. D. D. Bowman, 18–61. Alexandria, VA: WEF Press.

Stern, G. 1974. Pasteurization of liquid digested sludge. In Proceedings of the National Conference on Composting Municipal Sludge Management. Silver Spring, MD: Information Transfer, Inc.

Strauch, D. 1991. Survival of pathogenic micro-organism and parasites in excreta, manure, and sewage sludge. *Rev. Sci. Technol. Off. Int. Epizoot.* 10:813–46.

Tauxe, R. V. 1997. Emerging foodborne diseases: An evolving public health challenge. *Emerging Infect. Dis.* 3(4):425–34.

USDA. 1994. *Escherichia coli O157:H7 issues and ramifications*. Fort Collins, CO: USDA:APHIS:VS Centers for Epidemiology and Animal Health.

USEPA. 1994. *A plain English guide to the EPA Part 503 Biosolids Rule*. Report EPA/832/R-93-003. Washington, DC: U.S. Environmental Protection Agency, Office of Wastewater Management.

USEPA. 1999. *Control of pathogens and vector attraction in sewage sludge*. Cincinnati, OH: U.S. Environmental Protection Agency, Office of Research and Development, National Risk Management Research Laboratory, Center for Environmental Research Information.

USEPA. 2003. *Environmental regulations and technology control of pathogens and vector attraction in sewage sludge under 40 CFR Part 503*. EPA/625/R-92/013. Cincinnati, OH: U.S. Environmental Protection Agency.

Valcour, J. E., Michel, P., McEwen, S. A., and Wilson, J. B. 2002. Associations between indicators of livestock farming intensity and incidences of human shiga toxin-producing *Escherichia coli* infection. *Emerging Infect. Dis.* 8(3):252–57.

van Renterghem, B., Huysman, F., Rygole, R., and Verstraete, W. 1991. Detection and prevalence of *Listeria monocytogenes* in the agricultural system. *J. Appl. Bacteriol.* 71:211–17.

Walke, R. 1975. The preparation, characterization, and agricultural use of bark-sewage compost. PhD dissertation, University of New Hampshire, Durham.

Walker, F. R., and Stedinger, J. R. 1999. Fate and transport of *Cryptosporidium*. *J. Environ. Eng. ASCE.* 125(4):325–33.

Ward, R. L., Berstein, D. I., et al. 1986. Human rotavirus studies in volunteers: Determination of infective dose and serological response to infection. *J. Infect. Dis.* 154(5):871–80.

Ward, R. L., and Brandon, J. R. 1977. Effect of heat on pathogenic organisms found in wastewater sludge. In *National Conference on Composting Municipal Residues and Sludges*, 122–34. Rockville, MD: Information Transfer.

Wolfe, M. S. 1992. Giardiasis. *Clin. Microbiol. Rev.* 5:93–100.

Zhao, T., Doyle, M. P., Shere, J., and Garber, L. 1995. Prevalence of enterohemorrhagic *Escherichia coli* O157:H7 in a survey of dairy herds. *Appl. Environ. Microbiol.* 61(4):1290–93.

13 Bioaerosols

INTRODUCTION

Bioaerosols are organisms or biological agents that can be dispersed through the air and affect human health. Bioaerosols can contain living organisms, including bacteria, viruses, fungi, actinomycetes, arthropods, and protozoa, as well as microbial products such as endotoxins, microbial enzymes, β-1(→3)-glucans, and mycotoxins. Many of the bioaerosols are ubiquitous in the environment and may be affected by human activities (ACGIH, 1999). Bioaerosols are found in indoor and outdoor air environments. Organic matter, soil, plants, animals, and humans are the source of many bioaerosols.

Although bioaerosols can range from <0.01 μ to greater than 100 μ, they typically measure less than 20 μ (ACGIH, 1999; Goyer and Lavoie, 1998; Goyer et al., 2001). Particles >10 μm are filtered by hairs in the nose. Particles less <10 μm enter the lungs. Smaller particles, <5 μm, are respirable and can penetrate into the alveoli (lung air sacs), causing respiratory illnesses (Cole et al., 1999) (Figure 13.1).

Occupational exposure to bioaerosols occurs in numerous agricultural industries and farming enterprises that deal in residuals. These include:

- Farmers involved with animal manure, concentrated animal feeding operations (CAFOs), grain, etc.
- Composters involved with organic matter, yard materials, food waste, municipal solid waste, biosolids, etc.
- Workers in biosolids management, such as land application, composting, heat drying, lime stabilization
- Pulp and paper industry, wood products, and lumber workers
- Employees in the horticultural industry, e.g., greenhouses, landscaping, turf, and sod production
- Workers in zoological gardens
- Public works employees in parks and recreational facilities
- Veterinarians

In the United States asthma and allergies have been on the increase (Johanning, 1999a). Johanning (1999a) stated that in the period from 1980 to 1987, there was a 29% increase in the number of recorded cases of asthma in the United States, and there was an increase in asthma-related deaths. He estimated that the cost of treating these patients, as well as related costs, exceeded $6.5 billion/year. In 2002, it was estimated that 30.8 million persons in the United States had asthma (CDC, 2006).

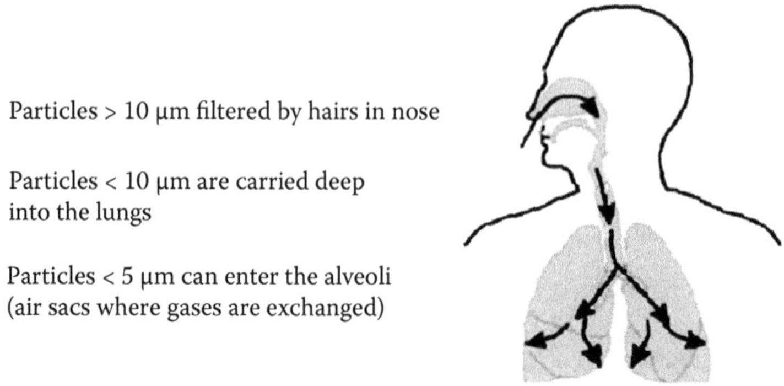

FIGURE 13.1 Particle sizes entering the lungs.

Asthma, or occupational or environmental type asthma, may be caused or associated with fungi and microbial by-products (Johanning, 1999a).

There have been more legal actions against composting operations related to bioaerosols and odors than other issues. Typically, residents experiencing odors fear that the air contains bioaerosols, which could impair their health. As a result of experiencing odors and believing that odorous air contains harmful agents such as bioaerosols, residents instigate lawsuits.

The major objectives of this chapter are to:

- Update the chapter on bioaerosols in the book *The Science of Composting* (Epstein, 1997)
- Provide the reader with information on the significance of bioaerosols
- Give information on the presence of bioaerosols in the environment
- Cover bioaerosols and composting
- Detail managing composting operations to reduce dispersion of bioaerosols

The following are some comprehensive reports, books, and documents on bioaerosols:

Rylander, R., and Jacobs, R. R., *Organic Dusts*, Lewis Publ., Boca Raton, FL, 1994

Cox, S., and Wathes, C. M., eds., *Bioaerosols Handbook*, Lewis Publ., CRC Press, Boca Raton, FL, 1995

Aspergillosis, Shering-Plough Corp., 2006

Prasad, M. et al., *Bioaerosols and Composting—A Literature Evaluation*, prepared by Cré members, Composting Association of Ireland TEO, 2004

Douwes, J., Thorne, P., Pearce, M., and Heedrik, D., Bioaerosol Health Effects and Exposure Assessment: Progress and Prospects, *Ann. Occup. Hyg.*, 47(3), 187–200, 2003

Latgé, J.-P., *Aspergillus fumigatus*, *Clin. Microb. Rev.*, 12(2), 310–50, 1999

Health and Safety Executive, *Occupational and Environmental Exposure to Bioaerosols from Compost and Potential Health Effects—A Critical Review of Published Data*, Research Report 130, prepared by the Composting Association and Health and Safety Laboratory, UK

BIOAEROSOLS ASSOCIATED WITH COMPOSTING AND THEIR POTENTIAL DISEASES

The bioaerosols discussed here are arbitrarily divided into two groups: major and minor. These two groups refer to their importance in transmitting diseases, primarily as related to residuals management. These will be discussed in some detail.

- Fungi
- *Aspergillus fumigatus*
- Endotoxin/organic dust
- Glucans
- Actinomycetes
- Mycotoxins

Fungi

Fungi as related to diseases are often cited as opportunistic organisms or secondary pathogens. Fungi are eukaryotic organisms (i.e., each fungal cell has at least one nucleus and nuclear membrane, endoplasmic reticulum, and mitochondria). Fungal cells resemble those of higher plants and animals. However, unlike plants, they do not have the ability to photosynthesize. They are common in soil, water, and decaying organic matter. Most fungi are obligate or facultative aerobic organisms and obtain nutrients from chemicals found in nature (Joklik et al., 1992).

Diseases, as related to exposure to fungi, generally cause infection, allergy, and toxicity (Johanning, 1999b). Immunocompromised patients are the most vulnerable to fungal infections. The most common pathogenic fungi are *Candida* spp., *Aspergillus* spp., and *Cryptococcus neoformans*. *Candida* spp. constitute the third to fourth most common cause of bloodstream infections occurring in hospitals or infirmaries (Walsh et al., 2004). *Aspergillus* spp. is the most common cause of pneumonic mortality in transplant patients (Walsh et al., 2004; Cunha et al., 1995). *Aspergillus fumigatus* is the primary organism. Rhodes (2006) states that *Aspergillus fumigatus* is the leading mold pathogen among immunocompromised patients, especially bone marrow and solid organ transplant patients. *Cryptococcus neoformans* is the most common cause of fungal-related mortality in HIV patients (Walsh et al., 2004), although candidiasis caused by *Candida albicans* is the most frequently identified common causative organism (Koll and Pawlecki, 1995). Walsh et al. (2004) indicate that there are less common but emerging fungal pathogens that cause morbidity and mortality in an increasing and expanding immunocompromised patient population. They indicate that *Fusarium* spp., *Scedosporium* spp., *Trichoderma* spp., *Zygoycetes,*

TABLE 13.1
Pathogenic Fungi and Their Environmental Sources

Fungus and Class of Fungus	Environmental Source	Reference
Aspergillus spp.	Soil, plant debris, organic matter, compost, air	Latgé, 1999; www.doctorfungus.org, 2006
Candida spp.	Gastrointestinal tract; leaves, flowers, water, and soil	Dagnani et al., 2003b www.doctorfungus.org, 2006
Cryptococcus neoformans	Soil contamination from avian excreta, pigeon droppings; decaying wood	Viviani et al., 2003 www.doctorfungus.org, 2006
Fusarium spp.	Common soil saprophyte; plants and soil; food, especially rice, bean, soybean, and small grains	Dignani et al., 2003a www.doctorfungus.org, 2006
Penicillium spp.	Soil, decaying vegetation, air	www.doctorfungus.org, 2006
Scedosporium spp.	Soil, polluted water, compost, manure	Dignani et al., 2003a
Trichoderma spp.	Soil, plant material, decaying vegetation, wood	www.doctorfungus.org, 2006
Trichosporon spp.	Soil, water, vegetables, mammals, birds, mouth, skin, nails	Maenza, and Merz, 2003; www.doctorfungus.org, 2006
Zygomycetes spp.	Decaying vegetables, foodstuff, fruits, soil, and animal excreta	Dromer and McGinnis, 2003

Penicillium marneffei, and *Trichosporons* spp. are the most significant emerging fungi. Table 13.1 shows the fungi or fungal class and environmental source. In Table 13.2 the diseases caused by the fungi as listed in Table 13.1 are shown. These diseases primarily affect immunocompromised individuals.

Bioaerosols are very common. Some, like *Aspergillus fumigatus*, are ubiquitous. Table 13.3 provides concentrations of bioaerosol fungi and bacteria in several activities and industries.

Aspergillus fumigatus

Aspergillus fumigatus is one of the ubiquitous organisms in the environment. It is a saprophytic fungus. It is associated with decaying organic matter, but is also present in food, waste, water, and soil. Figure 13.2 shows colonies of *Aspergillus fumigatus*. In sampling for *Aspergillus fumigatus* at composting sites, we typically use an Anderson microbial impactor. This unit collects the spores onto agar plates. Incubating the ager plate at thermophilic temperature destroys other microorganisms and preserves *Aspergillus fumigatus* since it is thermotolerant. In this manner, it is possible to count viable colonies. Knowledge of the volume of air sampled provides information on colonies per cubic meter of air (CFU/m^3).

Although we inhale hundreds of conidia daily, we are not infected with the organism. Inhalation of conidia by immunocompromised and immunocompetent

TABLE 13.2
Fungal Diseases

Fungus	Principal Organism	Disease(s)	References
Aspergillus spp.	*Aspergillus fumigatus*	Aspergillosis; invasive infection; allergic bronchopulmonary aspergillosis; toxicoses	Latgé, 1999; Walsh et al., 2004; Cunha and Meril, 1995; Andriole, 1993
Candida spp.	*Candida albicans*	Opportunistic mycosis; bloodstream infection; candidiasis of skin, mucosa, or nails; infections of mouth, vagina, nails, and skin It has been reported as the most frequent diagnosed infection in persons with HIV	Joklik et al., 1992; Walsh et al., 2004; Venes, 1997; www.doctorfungus.org, 2006; Koll and Pawlecki, 1995
Cryptococcus neoformans	*Cryptococcus neoformans*	Cryptococcosis and may involve skin, lungs, prostate gland, urinary tract, eyes, myocardium, bones, and joints Cryptococcal infections have been reported to be the most common life-threatening infections in AIDS patients	www.doctorfungus.org, 2006
Fusarium spp.	*F. solani*	Mycotoxicosis; infection in neutropenic patients and those undergoing transplantation	Dagnani et al., 2003a; www.doctorfungus.org, 2006
Penicillium spp.	*P. marneffei*	Penicillosis; pulmonary infection; lymphatic system; liver, spleen, and bones	www.doctorfungus.org, 2006; Dagnani et al., 2003a
Scedosporium spp.	*S. apiospermum* and *S. prolificans*	Invasive tissue disease; subcutaneous infections; mycetoma; osteomyelitis	Dagnani et al., 2003a; www.doctorfungus.org, 2006; Walsh et al., 2004
Trichoderma spp.	Five species have been identified in human diseases	Infections in immunocompromised individuals	Walsh et al., 2004
Trichosporon spp.	*T. asahi*, *T. beigelii*, and *T. mucoides*	Disseminated infection; white piedra	Walsh et al., 2004; www.doctorfungus.org, 2006
Zygomycetes spp.	Entomophthorales Mucorales	Rhinocerebral infection in diabetic patients; infection in immunocompromised individuals	Walsh et al., 2004

TABLE 13.3
Fungi and Bacteria Concentrations in Several Activities and Industries

Industry/Activity	Fungi (CFU/m^2)	Bacteria (CFU/m^2)
Agricultural harvesting and storage	10^3–10^9	10^2–10^3
Animal facilities	10^2–10^8	10^3–10^5
Composting	10^2–10^7	10^3–10^6
Manufacturing technology	10^2–10^6	10^2–10^6
Sawmill	10^4–10^8	10–10^3
Wastewater treatment (activated sludge)	10–10^3	10^2–10^6

Source: Prasad, M. et al., *Bioaerosols and Composting*, prepared by Cré members, Composting Association of Ireland TEO, Dublin, Ireland, 2004.

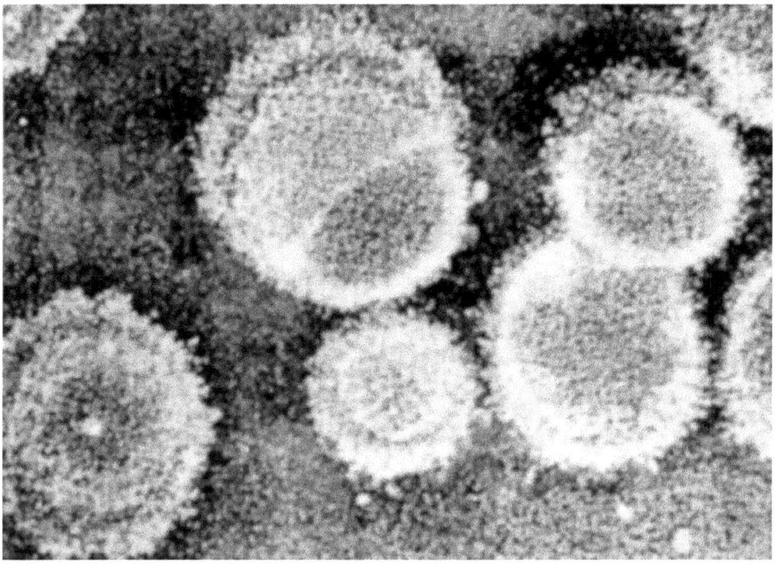

FIGURE 13.2 Colonies of *Aspergillus fumigatus* on oxgall agar, 45°C, twenty-four to forty-eight hours. (Courtesy of Dr. P. Millner, USDA, ARS, Beltsville, MD.)

individuals can have an adverse effect (Latgé, 1999; Hohl and Feldmesser, 2007). In competent humans, the conidia are killed and cleared by cells of the pulmonary immune system (Hohl and Feldmesser, 2007). In immunocompromised individuals, *Aspergillus fumigatus* can invade and infect the individual. This can occur in cancer patients, organ transplant patients, AIDS patients, and other highly debilitated patients (Cunha and Meril, 1995). The *Aspergillus* infection can result in morbidity. There is no infective dose for *Aspergillus fumigatus.* The clinical importance of invasive *Aspergillus* has increased in recent years, probably the result of increased immunocompromised persons. The main disease associated with this fungus is aspergillosis. The mortality rate of not treating this disease is nearly 100%. This disease accounts for up to 7% of deaths in Europe (Denning, 2006). Aspergillosis is a respiratory ailment. However, the organism can invade through surgical wounds or contaminated intravenous catheters.

The conidia that are released into the atmosphere are small (<3 μm) and can reach into the lung's alveoli (Latgé, 1999). It has been reported that all humans may inhale at least several hundred *Aspergillus fumigatus* conidia per day (Goodley et al., 1994; Latgé, 1999). It rarely causes significant disease in healthy individuals (Koll and Pawlecki, 1995). The most susceptible individuals are ones that are immunosuppressed or immunocompetent. However, Latgé (1999) indicates that in the period 1989 to 1999 *Aspergillus fumigatus* had become the most prevalent airborne fungal pathogen, causing severe and usually fatal invasive infections in immunocompromised hosts in developed countries. It has been reported more frequently in patients having an advanced stage of AIDS (Koll and Pawlecki, 1995).

Endotoxin

Endotoxin is a lipopolysaccharide (LPS) that is part of the cell wall of gram-negative bacteria. Rylander and Jacobs (1994) indicate that endotoxins are made up of complex LPS compounds, which consist of polysaccharide chains connected by a core oligosaccharide to a lipid part. Endotoxins are relatively heat stable. They are released into the environment during cell growth and after the cell dies, when the integrity of the cell wall is ruptured (Bradley, 1979). They are very common in the human and animal gut.

Endotoxins can be toxic to humans and animals. Endotoxins can cause fever and malaise (influenza-like symptoms). Inhaled endotoxins increase the activity of macrophages, which lead to a series of inflammatory conditions (Millner et al., 1994; Rylander, 2002; Milton, 1995; Milton et al., 1994). Rylander (2002) indicates that the internalization of endotoxin in macrophages and endothelial cells results in local production of inflammatory cytokines. The result is the migration of inflammatory cells into the lung and the penetration of cytokines into the blood. These lead to inflammation, toxic pneumonitis, and systemic symptoms.

Endotoxins present in organic dust have been implicated in toxic pneumonitis or organic dust toxic syndrome (ODTS) (Rylander et al., 1989; Rylander, 1994; NIOSH, 2002; Douwes et al., 2003). Inhalation of endotoxins can also result in a decrease in lung function and inflammatory responses (Michel et al., 1997).

Recently, it has been indicated that exposure to organic dust containing endotoxins may decrease the risk for atopic sensitization among children and decrease lung cancer among workers (Rylander, 2002). In the past, endotoxin health effects have been primarily associated with inducing airflow obstruction and aggravating asthma and allergies. However, it appears that endotoxins may play a beneficial secondary role since in early childhood they may have an atopy-protective effect by augmenting early Th-1 type immune development (Liu and Redmon, 2001).

Lange (2000) indicates that there is another important issue: occupational exposure to endotoxins may result in reduced lung cancer. This reduced lung cancer has been identified in textile, agriculture, and other workers (Enterline et al., 1985; Rylander, 1990; Mastrangelo et al., 1996; Lange, 2000).

GLUCANS

Glucans are ubiquitous. Fungal cell walls contain the polymer β-1(→3)-glucan, which is a polysaccharide composed of glucose units joined by β-1(→3)-glucan linkages. It is found in the cell walls of fungi, some bacteria, and cereals (barley and oats) (Millner et al., 1994; Williams, 1994). In the lung, β-1(→3)-glucan stimulate macrophages and neutrophils (ACGIH, 1999). Exposure to β-1(→3)-glucans has been attributed to respiratory impairment (Douwes et al., 1996). There is a limited amount of data on their presence in organic dust (Rylander et al., 1989). There is no specific data that connect the presence of airborne β-1(→3)-glucans, organic dust, and different effects occurring after exposure to organic dust. As Douwes et al. (2003) indicate, several small field studies have been performed in the home environment, daycare centers, office buildings, and schools, at household waste collectors, and with paper mill workers, suggesting a relation with respiratory symptoms, airway inflammation, lung function, and atopy in exposed individuals.

ACTINOMYCETES

Actinomycetes comprise a large and diverse group of gram-positive bacilli with a characteristic tendency to form chains or filaments. They are commonly found associated with soil and plant materials (Swan et al., 2003). The associated disease is actinomycosis. Actinomycetes are abundant in numerous agricultural and waste activities, such as mushroom and municipal waste composting, hay and grains processing, and processing plant fibers (sugarcane bagasse) (Millner, 1982; Lacey, 1990). Thermophilic actinomycetes thrive at temperatures in the range of 86°F (30°C) to 140°F (60°C). Millner (1982) and Swan et al. (2003) discuss the various species associated with composting. Lacey (1990) and Swan et al. (2003) indicate that actinomycetes are associated with allergies and respiratory diseases such as farmer's lung and mushroom worker's lung diseases. These are a form of extrinsic allergic alveolitis (Swan et al., 2002).

MYCOTOXINS

Mycotoxins produced by fungi are ubiquitous. Mycotoxins are low molecular weight, heat-stable, nonpolar compounds produced by fungi that could be toxic to humans and animals. Often they are found in food or animal feed and can cause disease or death when ingested. They are common in grain. The toxins can result in acute or chronic diseases that are neurotoxic, cytotoxic, mutagenic, teratogenic, or carcinogenic (Ciegler et al., 1981). One of the most noted mycotoxins in agriculture is aflatoxin produced by *Aspergillus flavus*. Other fungi, such as *Penicillium*, *Alternaria*, *Aspergillus* spp., *Fusarium*, *Stachybotrys*, and *Cladosporium* spp., produce mycotoxins that may have health effects (ACGIH, 1999). They are potent cytotoxins that cause cell disruption and interfere with essential cellular processes (ACGIH, 1999). Douwes et al. (2003) indicates that very little is known about occupational airborne exposure to mycotoxins and respiratory health effects.

Organic Dust

Organic dust can result in a disease termed organic dust toxic syndrome (ODTS). Symptoms can include fever, chest tightness, cough, and muscle aching. This is an acute respiratory and systemic illness that may affect workers in agriculture or those involved in residuals management (NIOSH, 2002). It is an acute, febrile, nonallergic, noninfectious respiratory illness (Madelin and Madelin, 1995). It is the result of organic dusts contaminated by microorganisms and microbial components, such as endotoxins, gram-negative bacteria, thermophilic microorganisms, and fungi (Schenker et al., 1991; Clark, 1994; Norn, 1994; Miller, 1994; Millner et al., 1994; Wintermeyer et al., 1997; ACGIH, 1999).

Organic dust is a generic term that includes all of the following conditions (NIOSH, 2002):

- Precipitin-negative farmer's lung disease
- Pulmonary mycotoxicosis
- Grain fever in grain elevators
- Silo unloader's syndrome
- Mill fever in cotton textile workers
- Inhalation fever

PREDOMINANT BIOAEROSOLS RELATED TO COMPOSTING AND THEIR PRESENCE IN THE ENVIRONMENT

Aspergillus fumigatus

There have been numerous studies assessing the prevalence of *Aspergillus fumigatus* in homes and hospitals (Solomon, 1976). Several papers have assessed the prevalence of *Aspergillus fumigatus* in outdoor air and the environment (Solomon, 1975; Baxter and Cookson, 1983; Hirsh and Sosman, 1976; Lumpkins et al., 1973; Kodama and McGee, 1986; Nobel and Clayton, 1963; Summerbell et al., 1989). These early data are discussed in my book *The Science of Composting*.

Workers in certain agricultural areas can be exposed to very high levels of bioaerosols such as *Aspergillus fumigatus*. Background levels can be very variable, with most being under 100 CFU/m^2. However, dusty areas, such as attics, can have high levels of *Aspergillus fumigatus*. Basements and moldy environments in the home have also been identified as sources of *Aspergillus fumigatus*.

Endotoxins

There are no regulations for endotoxins in the United States or Europe. There are suggested guidelines. The only guidance is from the International Commission on Occupational Health, which has published levels of endotoxin exposure that produce acute effects:

- 20–50 ng/m^3—mucous membrane irritation
- 100–200 ng/m^3—acute broncho constriction
- 1,000–2,000 ng/m^3—organic dust toxic syndrome

In the Netherlands, the Exposure Standards Setting Committee (DECOS) has proposed a personal, inhalable dust exposure measured as an eight-hour, time-weighted average of 200 EU/m^3 (Swan et al., 2003).

Endotoxins have been found in dust in homes (Park et al., 2000; Douwes et al., 2003). House dust endotoxins were associated with air conditioning (Gereda et al., 2001). Also, since endotoxins are common in house dust and homes with pets, especially with cats, infants and children, especially those with allergies or asthma, are at risk of endotoxin inflammation (Gereda et al, 2001; Park et al., 2000, 2004).

Table 13.4 shows levels of endotoxins reported for several agricultural and waste treatment activities. From this table it appears that workers in animal barns and farmers handling grain can be exposed to high concentrations of endotoxins. This high exposure to endotoxins could result in organic dust toxic syndrome (ODTS). All other reported workers in agricultural activities indicated in Table 13.4 can be exposed to endotoxin levels resulting in respiratory diseases.

Wouters et al. (2006) showed that domestic waste collectors can be exposed to high levels of endotoxins and inhalable dust. They reported that the range of endotoxins was <4 to 7,182 EU/m^3 (10 EU = 1 ng), and inhalable dust ranged from <0.2 to 9.1 mg/m^3. Heldal et al. (2003), in a very limited study, indicated that moderate exposure to fungal spores, endotoxins, and β-1(→3)-glucan during waste handling can induce respiratory inflammation. Endotoxins can be found during the processing of cotton, poultry, municipal solid waste, biosolids, bagasse, hemp, hay, grain, and vegetable dust (Rylander and Vesterlund, 1982; do Pico, 1986; Rylander and Jacobs, 1994; NIOSH, 2002).

TABLE 13.4
Endotoxin Levels for Several Agricultural and Waste Treatment Industries

Agricultural/Waste Treatment Activities	Endotoxin Levels ng/m^3	Reference
Municipal sewage workers	0–370	Melbostad et al., 1994
Cattle sheds	1,000–10,000	Swan et al., 2003
MSW recycling	1,000	Swan et al., 2003
Grain handling	Up to 70,000	Swan et al., 2003
Waste water treatment	Up to 300	Liesivuori et al., 1994
Indoor composting of biosolids	0–462	Epstein et al., 1997

In 1994 and updated in 2002, the National Institute of Occupational Safety and Health, Department of Health and Human Services, in NIOSH Publication 94-102, reported that certain agricultural workers and others may be exposed to organic dust resulting in ODTS. They recommended certain precautions to minimize risk.

These data clearly show that many bioaerosols such as *Aspergillus fumigatus* and endotoxins, as well as many other fungi, are common in the environment. We are constantly inhaling spores of these organisms. Latgé (1999) reported that environmental surveys show that all humans will inhale at least several hundred *Aspergillus fumigatus* conidia per day. Workers in certain agricultural and waste-handling industries may be exposed to high levels of bioaerosols. This could result in respiratory diseases.

Organic Dust

Organic dust, when present in high concentrations, can result in the disease ODTS. Thus, it has been identified in the agricultural environment (do Pico, 1986; Rylander and Jacobs, 1994; Paky and Knoblauch, 1995; NIOSH, 2002). NIOSH (2002) reported on two cases involving handling of wood chips and compost. Although there have not been any specific publications relating composting to ODTS, there have been health complaints and potential bioaerosol health implications (Epstein, 1997; Bünger et al., 2000; Heldal et al., 2003; Wouters et al., 2006).

BIOAEROSOLS AND COMPOSTING

Description of Bioaerosols Associated with Composting Operations

The two most important bioaerosols associated with composting are *Aspergillus fumigatus* and endotoxins. Organic dust, although it is not per se a bioaerosol, does contain a host of bioaerosols.

Aspergillus fumigatus

There are about three hundred *Aspergillus* species, of which *Aspergillus fumigatus* is the most pathogenic. This fungus has received the greatest attention with respect to residuals management. The main reasons are that it is ubiquitous, associated with decaying organic matter and soil, and thermotolerant. Growth can occur at temperatures as high as 55°C (131°F) and survive at temperatures up to 70°C (158°F) (Latgé, 1999). This is why it is not destroyed during composting. *A. fumigatus* is saprophytic and is important in recycling environmental carbon and nitrogen (Pitt, 1994). *Aspergillus fumigatus* is able to grow at elevated temperatures and utilize numerous varied sources of both carbon and nitrogen to support its growth. Therefore, it is an important opportunistic pathogen of humans, as well as a vital part of the nutrient recycling ecosystem (Rhodes, 2006). In composting, *A. fumigatus* is important in carbon and nitrogen recycling, primarily from nonwoody plant material (Tekaia and Latgé, 2005). The ability to sense carbon and nitrogen and to have the physiologic versatility to utilize different compounds contributes to the ability of this organism to compete with other fungi and bacteria in compost.

Endotoxins and Organic Dust

Endotoxins are ubiquitous and are found in organic dust. Consequently, organic dust can occur when the compost moisture is under 40%. Furthermore, organic dust can be generated because of spillage on roads and other areas of a site.

Measuring Bioaerosols at Composting Facilities

There are many ways to measure bioaerosols and dust. The measurement of bioaerosols at composting operations in the United States generally involves two methods: (1) Andersen microbial impactor, and (2) Bukard sampler. The Andersen impactor measures the viable organisms, such as *Aspergillus fumigatus*. Viable organisms can be responsible for causing disease, principally in immunocompromised persons. The disadvantage to this sampler is that it samples a specific point in time. Thus, numerous samples may need to be taken to characterize a potential condition. If, for example, one wants to determine conditions under various operations as related to potential for public or worker health, samples need to be taken during maximum dispersion conditions. The Bukard sampler requires microscopic examination and does not provide information on viability. It is basically like a sticky paper upon which organisms land. Its greatest disadvantage is that it does not provide information on the viability of the organisms. Therefore, the number of organisms, e.g., *Aspergillus fumigatus*, represents both viable (which cannot be distinguished) and nonviable ones. The latter may only cause allergic conditions. Furthermore, a competent biologist must examine the information under a microscope since some organisms are similar. Its greatest advantage is that one can get data over a long period of time. A combination of the two methods is ideal. This was done in the Islip, New York, study (NYDOH, 1994). However, this procedure is much more expensive.

In this author's experience, the use of the Andersen sampler during periods of maximum activity and dispersion has been well accepted in courts as an indication of the potential concentrations of *Aspergillus fumigatus*.

Personal dust samplers are used to conduct both respirable and total dust sampling. Components of a respirable dust sampler are a cyclone, a filter-cassette assembly, and a sampling pump. A total dust sampler does not have a cyclone. It contains a filter-cassette assembly and sampling pump. They can be used either for personal sampling or for activity sampling.

Endotoxins are typically measured using the *Limulus* amebocyte lysate (LAL) method. The test is simple, fast, and extremely sensitive.

Bioaerosols and Composting Operations

Bioaerosols may be generated during composting, while others, like *Aspergillus fumigatus*, are not destroyed by the composting process. The generation of bioaerosols and their potential dispersion depend on the composting system and operations.

Any system that agitates the compost, either with a specific machine or with a front-end loader, could result in the dispersion of bioaerosols. Prevention is discussed under the heading "Managing Bioaerosols" in this chapter. There have been numerous studies at various facilities to determine the level of bioaerosols.

Table 13.5 shows the concentration of *Aspergillus fumigatus* at several facilities in the United States and Europe.

Numerous studies have shown that bioaerosols emitted from composting operations reach background levels within 75 to 150 m (250 to 500 feet). One of the most comprehensive early documents was the result of a meeting of national and international experts (Millner et al., 1994). Some of the published data are shown in subsequent figures. Many of the examples were provided in my book *The Science of Composting* (Epstein, 1997). Figure 13.3 shows the dispersion of *Aspergillus fumigatus* from an open ASP site. At 150 m (500 feet) downwind during operations, the levels of *Aspergillus fumigatus* were at background levels.

Table 13.6 shows data on *Aspergillus fumigatus* at a municipal solid waste (MSW)/biosolids composting facility. Workers involved in certain areas need to protect themselves from exposure to high levels of bioaerosols. This is partcularly true in screening, tip floor, and other areas where agitation of the feedstock or compost is being done.

Schlosser et al. (2009) evaluated bioaerosols at six composting facilities in France, the UK, and Spain. Table 13.7 presents data on personal exposure to dust, endotoxins, mesophilic bacteria mesophilic molds, and actinomycetes using personal samples. They reported that mean exposure levels were from one hundred to ten thousand times higher than outdoor background levels. These levels were consistent with inflammatory and allergic respiratory effects to workers.

In a study in a biosolids composting facility in Longmont, Colorado, it was found that endotoxin levels correlated very well with total dust (Figure 13.4). In this facility, the screen was located within the composting building, and therefore workers were highly exposed to dust. Dust levels at times exceeded levels that could cause respiratory problems.

BIOAEROSOLS AND HUMAN HEALTH AS RELATED TO COMPOSTING

PUBLIC HEALTH

There is very little evidence that bioaerosols affect public health because of residuals management. Several reviews have been published, and most of the data related to levels of bioaerosols in various occupations (Dutkiewicz et al., 1988; Douwes et al., 2003). NRC (2002) reported that there was very little information on airborne pathogen occurrence during land application of biosolids. The primary concern is occupational exposure. Brooks et al. (2004) reviewed the emission, fate, and transport of bioaerosols from municipal and animal waste. They specifically evaluated potential exposure to the public in relation to land application of biosolids. They indicated that bioaerosols can be generated and released during wastewater treatment, land applied wastewater, land applied biosolids, and composting.

TABLE 13.5
Summary of Concentration of *Aspergillus fumigatus* at Various Composting Facilities

Location	Type of Composting Facility	Concentration (CFU/m^2)	Reference
Colorado	Biosolids, aerated static pile, enclosed	Mixing 1.1×10^3 Pile construction >74–77 $\times 10^2$ Pile breakdown 1.4 to >4.4×10^2 Pile screening <47 to 4.4×10^2 No activity 3 to 7	Epstein et al., 2001
Long Island, New York	Residential neighborhood near yard waste, outdoor	0.4×10^2 to 7.8×10^3	Recer et al., 2001
Norman, Oklahoma	MSW, outdoor	9.72×10^2	Folmsbee and Strevett, 1999
Maryland	Biosolids, enclosed	Upwind 0–34 (geo. mean 3.1) On-site 21–3,611(geo. mean 250) Downwind 0–30 (geo. mean 4.0)	Millner et al., 1994
Maine	Biosolids, outdoor	On-site 1–26 (mean 10) Downwind 30 m, 1 – 1,000 (mean 261)	Millner et al., 1994
Connecticut	Yard waste, outdoor	On-site 199 (mean) Upwind 4 (mean) Downwind 4–6 (mean 150–1,600 m)	Millner et al., 1994
New York	Yard waste	6×10^2	Millner et al., 1994
New Jersey	Biosolids	4×10^3 on-site 50 m downwind 1×10^3 250 m < 50 Control 0–2	Kothary et al., 1984
UK	Site 1: Botanic, kitchen Site 2: Green waste	Site 1: Turning 9×10^3 Site 2: Spreading 1.4×10^2	Gilbert et al., 2002
Italy	3 MSW facilities	1: 4.9×10^3 (these are maximum concentrations) 2: 2×10^2 3: 7.8×10	Varese et al., 2002
Germany	Enclosed system	2.03×10^3 near rotating screen 0.00 75 m upwind 3.00×10^2 downwind 7.77×10^1 control site	Danneberg et al., 1997
Germany	Literature	4.6×10^4 turning	Böhm et al., 2002
Windsor, Canada	Biosolids	2–1.15×10^2	Millner et al., 1994

Source: Prasad, M. et al., *Bioaerosols and Composting*, prepared by Cré members, Composting Association of Ireland TEO, Dublin, Ireland, 2004; Epstein, E. *The Science of Composting*, CRC Press, Boca Raton, FL, 1997; Millner, P. D. et al., *Compost Sci. Util.* 2(4), 6–57, 1994.

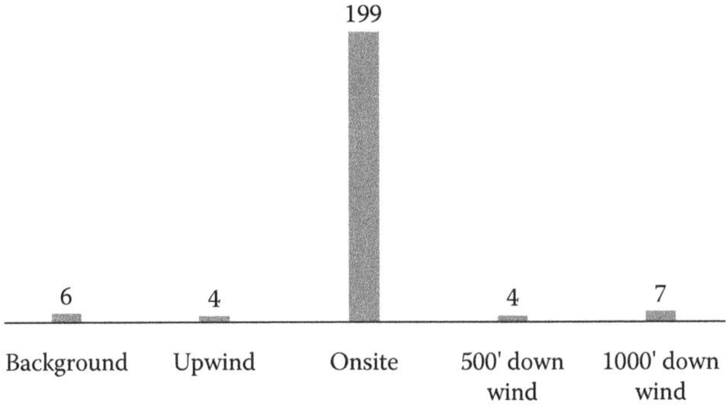

FIGURE 13.3 Dispersion of *Aspergillus fumigatus* at Site II, Maryland. (From Epstein, E. *The Science of Composting*, CRC Press, Boca Raton, FL, 1997.)

Major reviews and studies have been conducted in relation to the affect of bioaerosols as a result of composting (Millner et al., 1994; NYDOH, 1994; Epstein, 1997; Illinois Pollution Control Board, 1998; Swan et al., 2002, 2003). These studies included impacts of biosolids composting, yard waste composting, and municipal solid waste (MSW) composting. As a result, several governmental agencies specified minimal distances to residences. The Illinois Pollution Control Board, after an extensive review of the data, proposed to extend the 200 m (1/8 mile, 660 ft) setback from residences that was applied to composting areas, to also include health care facilities, preschool and child care facilities, and primary and secondary school facilities. The Environmental Agency (EA) in the United Kingdomhas established

TABLE 13.6
***Aspergillus fumigatus* at Various Locations in and around an MSW/Biosolids Composting Facility**

Sample Location	Concentration Range CFU/m²
Tip floor	857–11,714
Primary screen area	2,324–26,571
Compost floor during pile turning	>29,000
Postprocessing screen	>29,000
Outside building before working hours	2.4–5.9
Outside building during working hours	0–4.7
Property boundary—upwind	5.9–27.6
Property boundary—downwind	1.2–11.8
150 m downwind from property boundary	0–10.6
3,044 m downwind from property boundary	0–10.6
3.7 km downwind from property boundary	2.4–7.1

Source: Epstein, E. *The Science of Composting*, CRC Press, Boca Raton, FL, 1997.

TABLE 13.7
Geometric Mean of Personal Exposure to Bioaerosols as Related to Occupation

Task	Dust mg/m³	Endotoxin EU/m³	Mesophilic Bacteria cfu/m³	Mesophilic Molds cfu/m³	Actinomycetes cfu/m³
Inside cabin (n = 67)	0.9	2.1×10^3	1.5×10^6	7.9×10^4	5.4×10^4
Maintenance (n = 9)	3.6	5.7×10^4	3.2×10^7	7.5×10^4	1.2×10^5
Ground cleaning (n = 6)	2.5	2.4×10^4	1.0×10^8	2.6×10^5	3.0×10^5
Monitoring (n = 11)	2.1	8.0×10^3	4.5×10^6	7.3×10^4	1.3×10^5

Source: After Schlosser, O. et al., *Water Environ. Res.*, 81, 866–77, 2009.
Note: n = number of measurements.

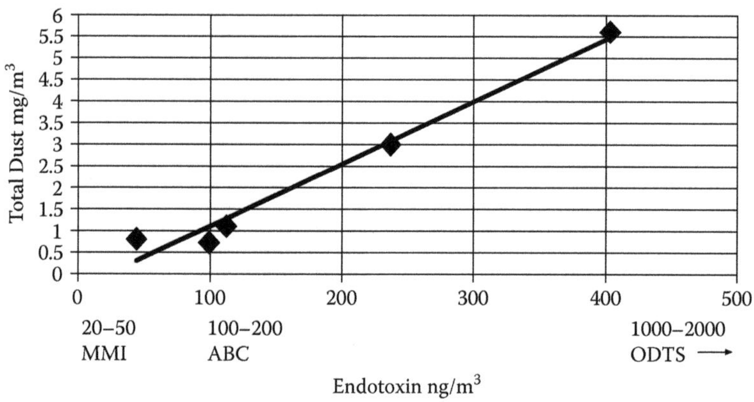

FIGURE 13.4 Endotoxin dispersion vs. total dust. (From Epstein, E. et al., *Compost Sci. Util.*, 9, 250–55, 2001.)

a policy regarding new composting facilities as follows: "Where the boundary of a facility is within 250 m (820 ft) of a workplace or the boundary of a dwelling, unless the application is accompanied by a site-specific risk assessment, based on clear independent scientific evidence" (EA 2001, 2008). A report to the Health and Safety Executive, a UK governmental agency, listed only two published case studies that showed evidence of respiratory infection, such as allergic bronchopulmonary aspergillosis—one from the United States and one from mainland Europe.

Worker Health

Workers are the most exposed individuals to bioaerosols. They are exposed to much higher concentrations than the public would ever experience. Furthermore, they are exposed for much longer periods of time than the public.

The most comprehensive worker evaluation was conducted from 1987 to 1991 at the biosolids and wood chip composting operation located in Site II in Maryland.

Blood samples, lung function tests, and other direct measurements were conducted (Chesapeake Occupational Health Services, 1991). The data are shown in the author's book *The Science of Composting* (Epstein, 1997, pp. 282–283). Reports of asthma, bronchitis, earaches, and shortness of breath were infrequent. Spirometric findings were also very low. There was no evidence of adverse worker health.

A study for the Canadian Ministry of Labour evaluating yard waste workers in Canada did not find any occupational health effects (Green Lane Environmental Group Ltd., 1995). It did recommend protective and hygienic measures. In one biosolids composting facility, this author encountered an individual that developed a skin rash. After receiving dermatological treatment, the rash disappeared.

Workers in the waste recycling industry involving organic waste sorting, collection, and composting are often exposed to very high levels of bioaerosols (Douwes et al., 2000). Bünger et al. (2000) reported that compost workers had significantly more symptoms and diseases of airways and skin than did control subjects. The facility handled biowaste (separated organic wastes). Manual sorting removed noncompostable materials. The above-mentioned authors indicated that until the time of the publication there was very little published information on health risks of compost workers. Bünger et al. (2000) found that exposure to organic dust at composting facilities is associated with adverse acute and chronic respiratory health effects. They evaluated workers at forty-one German composting facilities, which included biowaste and yard waste. Filamentous fungi and thermotolerant/thermophilic actinomycetes were at high concentrations. The health symptoms reported were mucosal membrane irritation of the eye, conjunctivitis, and in upper airways of the lungs.

Wouters et al. (2006) reviewed and reported on personal occupational exposure to inhalable dust, endotoxins, β(1-3)-glucan, and fungal extracellular polysaccharides in waste management operations. Their results showed that endotoxin and glucan exposure levels were relatively low for greenwaste composting. Levels were approximately 300 to 1,000 EU m^{-3} for endotoxin and 5 to 10 μg m^{-3} for glucan. Exposure variability was large. Occupational exposure limits for organic dust and endotoxin are frequently exceeded, suggesting that workers are at risk for developing adverse health effects.

Schlosser et al. (2009) reported on an association between occupational exposure to bioaerosols in composting facilities and health effects. They evaluated conditions in six plants. Sampling results showed large ranges of concentrations of dust, bacteria, molds, and endotoxins in ambient air and in personal samples, both when driving front-end loaders and when cleaning, monitoring, and performing maintenance tasks. Some of their data are shown in Table 13.8. The study included personal exposure and seasonal variation. Their paper also presented an excellent review.

Drew et al. (2009) evaluated the quality of bioaerosol risk assessment for composting facilities in England and Wales. The evaluation involved forty-four environmental risk assessments.

There are approximately fifteen MSW composting plants in the United States. At one U.S. solid waste and biosolids composting facility, this author encountered a front-end loader operator who felt better on Monday than on Friday. This indicated possible organic dust respiratory impairment. One of the problems was that he was

TABLE 13.8
Geometric Means of Bioaerosols in Air in Composting Facilities

Process Area	Dust mg/m³	Endotoxin EU/m³	*A. fumigatus* cfu/m³	Actinomycetes cfu/m³
Mixing—composting	3.6	1.2×10^5	1.8×10^4	2.4×10^5
Screening	7.5	1.3×10^5	8.0×10^3	5.5×10^5
Curing	1.1	2.3×10^4	4.0×10^3	1.4×10^5
Shredding	1.2	5.2×10^4	1.5×10^3	4.0×10^4
Background	0.3	76	16	2.9×10^3

Source: After Schlosser, O. et al., *Water Environ. Res.*, 81, 866–77, 2009.

a heavy smoker. To minimize any future lung problems due to bioaerosols, he was assigned to a new job where he was not exposed to dust.

In Edmonton, Canada, there exists the largest MSW composting plant in North America. This plant had considerable dust problems. To date there are no reports or documented evidence of occupational illnesses from this plant or the ones in the United States.

There have been considerable recent data coming from Europe on worker health. Conditions in Europe are very different from those in the United States. In the United States, there are very few MSW or biowaste composting facilities. Landfills are inexpensive for most communities. Composting of MSW would be more expensive. Most of the five thousand composting facilities in the United States are operated in the open, since land is inexpensive. In Europe, land is not as available, and composting of MSW or biowaste is in enclosed facilities. Enclosed facilities tend to have more dust, as seen in the Edmonton, Canada, MSW biosolids facility. However, worker health knowledge is useful. Under certain conditions, certain components in a facility could produce considerable dust. These could be feedstock delivery buildings and screening areas. Workers should be protected, and the facilities should have proper ventilation.

Heldal et al. (2003) assessed airway inflammation using induced-sputum analysis, analyzed for interleukin-8, measured eosinophil cationic protein, and took spirometric measurements of workers handling waste. Bünger et al. (2000) reported on a five-year follow-up study at forty-one composting facilities in Germany. A total of 218 workers and 66 control subjects were involved in the study. The study involved a questionnaire, a clinical examination, and spirometric measurements. They concluded that exposure to organic dust at workplaces of composting facilities is associated with adverse acute and chronic respiratory health effects, including mucosal membrane irritation (MMI), chronic bronchitis, and reduced forced vital capacity (FVC) of the lungs.

There have been several papers on MSW facilities in Europe that have examined worker health. Most of these have been in Scandinavia, Netherlands, and Germany (Nersting et al., 1991; Sigsgaard et al., 1990; Malmros, 1990; Krajewski et al., 2002; Heldal et al., 2003; Wouters et al., 2006; Harrison, 2007). Three texts published on bioaerosols (Rylander and Jacobs, 1994; Burge, 1995; ACGIH, 1999) provide very little information on the effect from bioaerosols to workers in the waste industry. Most of the data from Europe have been written with regard to measurement of bioaerosols rather than direct measurement and evaluation of workers.

MANAGING BIOAEROSOLS

Managing bioaerosols to minimize exposure to the public requires two aspects: (1) moisture control, and (2) dust control. Moisture control depends on whether the facility is enclosed or operated in the open. In either case, moisture control is important for proper biological management and decomposition. Enclosed systems must be designed to prevent excess moisture loss during the composting process. Moisture during the composting process should be in the range of 40 to 55%. Below 40%, excess dust can occur, and above 60%, the pore space in the media is blocked and oxygen is inhibited. An agitated bed technology needs to have a system for the addition of water during the composting phase. Other enclosed systems need to prevent excessive drying.

Open systems need to control moisture loss. Windrow systems need to control moisture during agitation. This can be done by having a watering system as part of the windrow equipment or having a water truck following the turner and spraying water. This will reduce both bioaerosol and dust dispersion. With ASP systems, the blower control and covers will reduce moisture loss. If during an ASP system the compost is being moved, e.g., from one bin or pile to another, moisture may need to be added. Several ASP systems will increase the blower time toward the end of the process to enable the compost to be screened. Most screening systems are most effective when the moisture content is approximately 40 to 45%. At less than 40%, dust can occur. Some facilities have installed a very fine spray system over the screening device to reduce dust. This spray does not increase the moisture content of the compost, nor does it affect the screening process. During curing, especially in large piles, very little moisture is lost except at the surface. Surface irrigation can reduce moisture loss. Some curing systems cover the piles with a fabric to reduce both moisture loss and dust.

CONCLUSION

Bioaerosols are ubiquitous. Dr. Rippon, former director of Mycology Service Laboratory, University of Chicago, Pritzker School of Medicine, stated: "We are all exposed to conidia of *Aspergillus fumigatus* and the 900 other species of *Aspergillus* on a daily basis." Many of these bioaerosols can affect human health, especially through the respiratory system. Several of the bioaerosols found in residuals or emitted during residuals processing and management have been shown to cause illnesses in the agricultural industry. Farmer's lung, mill fever (cotton processing), and grain fever (organic dust toxic syndrome) are just a few. Organic dust has been implicated in swine confinement buildings and wood processing (Rylander and Jacobs, 1994), as well as in workers shoveling wood chips (NIOSH, 2002).

There is considerable literature on the health effects of bioaerosols in the indoor environment (Burge, 1990, 1995; Epstein, 1997; Herr et al., 2003; Anaisse et al., 2003). Bioaerosol health effects to immunosuppressed individuals in hospitals are well documented (Cunha and Meril, 1995; Latgé, 1999; Anaisse et al., 2003). The majority of *Aspergillus fumigatus* infections reported in the medical literature have occurred in hospitals.

It is estimated that there are between thirty thousand and fifty thousand workers in the composting industry in the United States. In addition, there are many more workers in wastewater treatment and solid waste facilities. Compost workers are the most exposed individuals. They are exposed to much greater concentrations and much more frequently than the general public. There is no evidence in the medical literature that workers in the composting industry are prone to a higher incidence of disease than the general public. Similar data are not indicated for wastewater workers or workers in the municipal solid waste area. Neither the Centers for Disease Control (CDC), USEPA, or NIH has issued warnings or regulations regarding the public or workers in the residuals management field.

Workers in residuals need to take precautions and practice good personal hygiene. Workers exposed to dust should wear dust masks. The industry needs to minimize worker exposure to dust and bioaerosols through proper ventilation in enclosed facilities, as well as in mobile working equipment.

There is also a need for good epidemiological studies on worker health among employees in the waste and recycling industry. Most of the data have come from Europe and have been primarily showing levels of bioaerosols rather than specific human health. Most of the data reported were from the use of questionnaires rather than direct measurement of respiratory conditions or other medical examinations. One of the most difficult problems is the lack of dose-response relationships for various bioaerosols. Thus, the levels of bioaerosols that can inflict illnesses in humans are unknown.

REFERENCES

ACGIH. 1999. *Bioaerosols: Assessment and control.* Cincinnati, OH: ACGIH.

Aisner, J., Schimpff, S. C., Bennett, J. E., Young, V. M., and Wiernik, P. H. 1976. *Aspergillus* infections in cancer patients: Association with fireproofing materials in a new hospital. *JAMA* 235:411.

Anaissie, E. J., McGinnis, M. R., and Pfaller, M. A. 2003. *Clinical mycology*. New York: Churchill Livingston.

Andriole, V. T. 1993. Infections with *Aspergillus* species. *Clin. Infect. Dis.* 17:S481–86.

Baxter, L. J., and Cookson, J. T. 1983. Natural atmospheric microbial conditions in a typical suburban area. *Appl. Environ. Microbiol.* 45(3):919–34.

Bodey, G. P., and Vartibvarian, S. 1989. Aspergillosis. *Eur. J. Clin. Micrbiol. Infect. Dis.* 8(5):413–37.

Böhm, R. E., Martens, W., and Phillipp, W. 2002. Hygienic relevance of the extension of bacteria with the collection and treatment of wastes. Abfallforum online 22 as quoted by.

Bradley, S. G. 1979. Cellular and molecular mechanisms of action of bacterial endotoxins. *Am. Rev. Microbiol.* 33:67–94.

Brooks, J., Gerba, C. P., and Pepper, I. L. 2004. Biological aerosol emission, fate, and transport from municipal and animal wastes. *J. Residuals Sci. Technol.* 1(1):15–28.

Bünger, J., Antlauf-Lammers, M., Schulz, T. G., Westphal, G. A., Müller, M. M., Ruhnau, P., and Hallier, E. 2000. Health complaints and immunological markers of exposure to bioaerosols among biowaste collectors and compost workers. *Occup. Environ. Med.* 57:458–64.

Burge, H. A. 1990. Bioaerosols: Prevalence and health effects in the indoor environment. *J. Allergy Clin. Immunol.* 86:687.

Burge, H. A. 1995. *Bioaerosols*. Boca Raton, FL: Lewis Publishers.

CDC. 2006. *Asthma prevention, health care use, and mortality, 2002*. Centers for Disease Control, National Center for Health, U.S. Department of Health and Human Services, MedlinePlus.

Chesapeake Occupational Health Services. 1991. *Health surveillance program for compost workers: An epidemiological review*. Report to Washington Suburban Sanitary Commission. Silver Spring, MD: Chesapeake Occupational Health Services.

Ciegler, A., Burmeister, H. R., Wesonder, R. F., and Hesseltine, C. W. 1981. Mycotoxins: Occurrence in the environment. In *Mycotoxins and N-nitroso compounds: Environmental risks*. Vol 1. Boca Raton, FL: CRC Press, Inc.

Clark, C. S. 1994. Waste handling. In *Organic dusts*, ed. R. Rylander and R. R. Jacobs, 257–63. Boca Raton, FL: Lewis Publishers.

Cole, D., Hill, V. R., Humenick, F. J., and Sobsey, M. D. 1999. Health, safety, and environmental concerns of farm animal waste. *Occup. Med.* 14(2):423–48.

Cox, S., and Wathes, C. M., eds., 1995. *Bioaerosols Handbook*, Boca Raton, FL: Lewis Publ., CRC Press.

Cunha, B. A., and Meril, G. J. 1995. *Aspergillosis*. Special report. Montvale, NJ: Medical Economics.

Danneberg, G., Gruneklee, E., Seitz, M., Hartung, J., and Driesel, A. 1997. Microbial and endotoxin emissions in the neighborhood of a composting plant. *Ann. Agric. Environ. Med.* 4:169–73.

Dagnani, M. C., Kiwan, E. N., and Anaissie, E. J. 2003a. Hyalohyphomycoses. In *Clinical mycology*, ed. E. J. Anaissie, M. R. McGinnis, and M. A. Pfaller. New York: Churchill Livingston.

Dagnani, M. C. S., Solomkin, J. S., and Anaissie, E. J. 2003b. Candida. In *Clinical mycology*, ed. E. J. Anaissie, M. R. McGinnis, and M. A. Pfaller. New York: Churchill Livingston.

Denning, D. W. 2006. *Aspergillosis*. Manchester, UK: Education and Research Centre, Wythenshawe Hospital.

do Pico, G. A. 1986. Health effects of organic dusts in the farm environment. Report on diseases. *Am. J. Indus. Med.* 10:261–65.

Douwes, J., Doekes, G., Montijn, R., Heederik, D., and Brunekreef, B. 1996. Measurement of B(1→3)-glucans in occupational and home environments with an inhibition enzyme immunoassay. *Appl. Environ. Microbiol.* 1996:3176–82.

Douwes, J., Thorne, P., Pearce, N., and Heederik, D. 2003. Bioaerosol health effects and exposure assessment: Progress and prospects. *Ann. Occup. Hyg.* 47(3):187–200.

Drew, G. H., Jordinson, G. M., Smith, M. A., and Pollard, S. J. T. 2009. Evaluating the quality of bioaerosols risk assessments for composting facilities in England and Wales. *Resources Conservation Recycling* 53:507–12.

Dromer, F., and McGinnis, M. R. 2003. Zygomycosis. In *Clinical mycology*, ed. E. J. Anaissie, M. R. McGinnis, and M. A. Pfaller. New York: Churchill Livingston.

Dutkiewicz, J., Jablonski, L., and Olenchock, S. A. 1988. Occupational biohazards. A review. *Am. J. Ind. Med.* 14:605–23.

EA. 2001. Environmental Agency position on composting and health effects. Bristol 2001. Available at http://www.environment-agency.gov.uk.

EA. 2008. Environmental Agency sustainable management of biowastes. Composting—Maximizing the benefits and minimizing the environmental impacts. Environmental Agency position statement, Bristol 2008. Available at http://www.environment-agency.gov.uk.

Enterline, P. E., Sykora, J. L., Keleti, G., et al. 1985. Endotoxin, cotton dust, and cancer. *Lancet* ii:934–35.

Epstein, E., Wu, N., Youngberg, C., and Crouteau, G. 2001. Dust and bioaerosols at a biosolids composting facility. *Compost Sci. Util.* 9:250–55.

Epstein, E. 1997. *The science of composting*. Lancaster, PA: Technomic Publ. Co., Inc.

Epstein, E., Williams, T. O., Youngberg, C., and Wu, N. 1997. Dust and bioaerosol assessment at two biosolids composting facilities. In *WEFTEC '97*. Alexandria, VA: Water Environment Federation.

Folmsbee, M., and Strevett, K. A. 1999. Bioaerosol concentration at an outdoor composting center. *J. Air Waste Manage. Assoc.* 49:554–61.

Gereda, J. F., Klinnert, M. D., Price, M. R., Leung, D. Y., and Liu, A. H. 2001. Metropolitan home living conditions associated with indoor endotoxin levels. *J. Allergy Clin. Immunol.* 107(5):790–96.

Gilbert, J. E., Kelsey, A., Karnon, A., Swan, J. D., and Crook, B. 2002. *Preliminary results of monitoring the release of bioaerosols from composting facilities in the UK: Interpretation, modeling, and appraisal of mitigation measures. International Symposium on Composting and Compost*. Emmaus, PA: J.G. Press.

Goodley, J. M., Clayton, Y. M., and Hay, R. J. 1994. Environmental sampling for aspergilli during building construction on a hospital site. *J. Hosp. Infect.* 26:27–35.

Goyer, N., and Lavoie, J. 1998. *Emissions du traitement secondaire des effluents des papetieres*. Report R-2001. Montreal, Canada: Etudes et recherches, IRSST.

Goyer, N., Lavoie, J., Lazure, L., and Marchand, G. 2001. Les bioaérosols en milieu de travail: Guide d'évaluation, de contrôle et de prévention. Études et recherches. Technical Guide T-23. Montreal: IRSST.

Green Lane Environmental Group Ltd. 1995. Sarnia leaf and yard composting site bioaerosol control study. Sarnia, Lambeth, Ontario, Canada: Sarnia Leaf and Yard Composting Facility.

Harrison, E. Z. 2007. *Compost facilities: Off-site air emissions and health*. Ithaca, NY: Cornell Waste Management Institute.

Health and Safety Executive, *Occupational and Environmental Exposure to Bioaerosols from Compost and Potential Health Effects—A Critical Review of Published Data*, Research Report 130, prepared by the Composting Association and Health and Safety Laboratory, UK.

Heldal, K. K., Halstensen, A. S., Thorn, J., Djupesland, P., Wouters, I., Eduard, W., and Halestensen, T. S. 2003. Upper airway inflammation in waste handlers exposed to bioaerosols. *Occup. Environ. Med.* 60:444–50.

Herr, C. E. W., Nieden, A. Z., Jankofsky, M., Stilianakis, N. I., Boedeker, R.-H., and Eikmann, T. F. 2003. Effects of bioaerosol polluted outdoor air on airways of residents: A cross sectional study. *Occup. Environ. Med.* 60:336–42.

Hirsch, S. R., and Sosman, J. A. 1976. A one-year survey of mold growth inside twelve homes. *Ann. Allerg.* 36:30–38.

Hohl, T. M., and Feldmesser, M. 2007. *Aspergillus fumigatus:* Principles of pathogenesis and host defense. Downloaded from ec.asm.org at the John Rylands University Library, University of Manchester, UK.

Illinois Pollution Control Board. 1998. Amendments to Requirements for Landscaping Waste Compost Facilities, 35 ILL. ADM. CODE 830.203(c) 831.107, and 831.1099b0(3).

Johanning, E. 1999a. Fungi in indoor environments—A challenge for scientific research and public health. In *Bioaerosols, fungi, and mycotoxins: Health effects, assessment, prevention, and control*, ed. E. Johanning, 12–21. Albany: Eastern New York Occupational and Environmental Health Center.

Johanning, E. 1999b. Fungal and related exposure. In *Occupational medicine secrets*, ed. R. M. Bowler and J. E. Con, 135–46. Philadelphia, PA: Hanley & Belfus, Inc.

Joklik, W. K., Willett, H. P., Amos, D. B., and Wilfert, C. M., eds. 1992. *Zinsser microbiology*. 20th ed. Norwalk, CT: Appleton & Lange.
Kodama, A. M., and McGee, R. I. 1986. Airborne microbial contaminants in indoor environments: Naturally ventilated and air conditioned homes. *Arch. Environ. Health* 41(5):306–11.
Koll, B. S., and Pawlecki, J. B. 1995. *Fungal infection in the AIDS patient*. Montvale, NJ: Internal Medicine, Economic Medicine.
Kothary, M. H., Chase, T. J., and Macmillan, J. D. 1984. Levels of *Aspergillus fumigatus* in air and in compost at a sewage sludge composting site. *Environ. Pollut.* (Series A) 34:1–14.
Krajewski, J. A., Tarkowski, S., Cyprowski, M., Szarapinska-Kwaszewska, J., and Dudkiewicz, B. 2002. Occupational exposure to organic dust associated with municipal waste collection and management. *Int. J. Occup. Med. Environ. Health* 15(3):289–301.
Lacey, J. 1990. Aerobiology and health: The role of airborne fungi spores in respiratory disease. In *Frontiers in Mycology*, ed. D. L. Hawksworth. Kew, UK: C.A.B. International.
Lange, J. H. 2000. Reduced cancer rates in agricultural workers: A benefit of environmental and occupational endotoxin exposure. *Med. Hypotheses* 55:383–85.
Latgé, J.-P. 1999. *Aspergillus fumigatus* and aspergillosis. *Clin. Microbiol. Rev.* 12(2):210–350.
Liesivuori, J., Kotimaa, M., Laitinen, S., Louhelainen, K., Ponni, J., Sarantila, R., and Husman, K. 1994. Airborne endotoxin concentrations in different work conditions. *Am. J. Indust. Med.* 25:123–24.
Liu, A. H., and Redmon, A. H. J. 2001. Endotoxin: Friend or foe. *Allergy Asthma Proc.* 22(6):337–40.
Lumpkins Sr., E. D., Corbit, S. L., and Tiedeman, G. T. 1973. Airborne fungi survey. 1. Culture-plate survey of the home environment. *Ann. Allergy* 31:361–70.
Madelin, T. M., and Madelin, F. 1995. Biological analysis of fungi and associated molds. In *Bioaerosols handbook*, ed. C. S. Cox and C. M. Mathes. Boca Raton, FL: CRC Press, Inc.
Maenza, J. R., and Merz, W. G. 2003. Infections caused by non-candida, non-cryptococcus yeasts. In *Clinical mycology*, ed. E. J. Anaissie, M. R. McGinnis, and M. A. Pfaller. New York: Churchill Livingston.
Malmros, P. 1990. *Problems with the working environment in solid waste treatment*. Rep. 10/1990. Copenhagen: National Labour Inspection of Denmark.
Mastrangelo, G., Marzia, V., and Marcer, G. 1996. Reduced lung cancer mortality in dairy farmers; is endotoxin exposure the key factor? *Am. J. Ind. Med.* 30:601–9.
Melbostad, E., Eduard, W., Skogstad, A., Lassen, J., Sostrand, P., and Heldal, K. 1994. Exposure to bacterial aerosols and work-related symptoms in sewage workers. *Am. J. Indust. Med.* 25:59–63.
Michel, O., Gianni, R., Le Bon, B., Content, J., Duchateau, J., and Sergysels, J. 1997. Dose-response relationship to inhaled endotoxin in normal subjects. *Am. J. Resp. Crit. Care Med.* 145:1157–64.
Miller, J. D. 1994. Mycotoxins. In *Organic dusts: Exposure, effects, and prevention*, ed. R. Rylander and R. R. Jacobs, 87–92. Boca Raton, FL: Lewis Publ.
Millner, P. D. 1982. Thermophilic and thermotolerant actinomycetes in sewage-sludge compost. Symposium: Non-antibiotic producing actinomycetes. *Dev. Indust. Microbiol.* 23:61–78.
Millner, P. D., Olenchock, S. A., Epstein, E., Rylander, R., Haines, J., Walker, J., Ooi, B. L., Horne, E., and Maritato, M. 1994. Bioaerosols associated with composting facilities. *Compost Sci. Util.* 2:6–57.
Milton, D. K. 1995. Endotoxin. In *Bioaerosols*, ed. H. A. Burge, 77–87. Boca Raton, FL: Lewis Publ.

Milton, D. K., Kriebel, D., Wypij, D., Walters, M., Hammond, K., and Evans, J. S. 1994. Acute and chronic airflow obstruction and endotoxin exposure. *Am. J. Respir. Crit. Care Med.* 149:A399.

Nersting, L., Malmros, P., Sigsgaard, T., and Petersen, C. 1991. Biological health risk associated with resource recovery, sorting of recycled waste and composting. *Grana* 30:454–57.

NIOSH. 2002. Request for assistance in preventing organic dust toxic syndrome. Cincinnati, OH: Centers for Disease Control, National Institute of Occupational Safety and Health.

Nobel, W. C., and Clayton, Y. M. 1963. Fungi in the air of hospital wards. *J. Gen. Microb.* 32:397–402.

Norn, S. 1994. Microorganisms-induced or enhanced mediator release: A possible mechanism in organic dust related diseases. *Am. J. Indust. Med.* 25(1):91–95.

NRC. 2002. *Biosolids applied to land.* Washington, DC: National Research Council, National Academies Press.

NYDOH. 1994. A prospective study of health symptoms and bioaerosol levels near a yard waste composting facility. Islip Composting Facility, Town of Islip, Suffolk County, New York. Albany: State of New York, Department of Health.

Opal, S. M., Asp, A. A., Cannady Jr., P. B., Morse, P. L., Burton, L. J., and Hammer II, P. G. 1986. Efficacy of infection control measures during nosocomial outbreaks of disseminated aspergillosis associated with hospital construction. *J. Infect. Dis.* 153:634–37.

Paky, A., and Knoblauch, A. 1995. Dust exposure, dust-induced lung disease and respiratory protective systems in farming. *Schweizerische Medizinische Wochenschrift* 125(10):458–56.

Park, J. H., Spiegelman, D. L., Burge, H. A., Gold, D. R., Chew, G. L., and Milton, D. K. 2000. Longitudinal study of dust and airborne endotoxin in the home. *Environ. Health Perspect.* 108(11):1023–28.

Park, J. H., Szponar, B., Larsson, L., Gold, D. R., and Milton, D. K. 2004. Characterization of lipopolysaccharides present in settled house dust. *Appl. Environ. Microb.* 70(1):262–67.

Pitt, J. I. 1994. The current role of *Aspergillus* and *Penicillium* in human and animal health. *J. Med. Vet. Mycol.* S1:17–32.

Prasad, M. et al. 2004. *Bioaerosols and Composting*, prepared by Cré members, Composting Association of Ireland TEO, Dublin, Ireland.

Recer, G. M., Browne, N. O., Horn, E. G., Hill, K. M., and Boehler, W. F. 2001. Ambient air levels of *Aspergillus fumigatus* and thermophilic actinomycetes in a residential neighborhood near a yard-waste composting facility. *Aerobiology* 17:99–108.

Rhodes, J. C. 2006. *Aspergillus fumigatus*: Growth and virulence. *Med. Mycol.* 44:577–81.

Rodenhuis, S., Beaumont, F., Kauffman, H. F., and Sluiter, H. J. 1984. Invasive pulmonary aspergillosis in a non-immunosuppressed patient: Successful management with systemic amphotericin and flucytosine and inhaled amphotericin. *Thorax* 39:78–79.

Rylander, R. 1990. Environmental exposure with decreased risk for lung cancer. *Int. J. Epidemiol.* 19:56772.

Rylander, R. 1994. Endotoxins. In *Organic dusts: Exposure, effects, and prevention*, ed. R. Rylander and R. J. Jacobs. Boca Raton, FL: Lewis Publ.

Rylander, R. 2002. Endotoxin in the environment—Exposure and effects. *J. Endotoxin Res.* 8(4):241–52.

Rylander, R., Bergstrom, R., Goto, H., Yoasa, K., and Tanaka, S. 1989. Studies on endotoxin and (1-3)-β-D-glucan in cotton dust. In *Proceedings of the Thirteenth Special Session on Cotton Dust Research*, ed. R. R. Jacobs and P. J. Wakelyn, 46–47. Memphis, TN.

Rylander, R., and Jacobs, R. R. 1994. *Organic dust exposure, effects, and prevention.* Boca Raton, FL: Lewis Publ.

Rylander, R., and Vesterlund, J. 1982. Airborne endotoxin in various occupational environments. *Prog. Clin. Biol. Res.* 93:399–409.

Schenker, M., Ferguson, T., and Gamsky, T. 1991. Respiratory risks associated with agriculture. *Occup. Med. State Art Rev.* 6(3):415–28.

Schlosser, O., Huyard, A., Cartnick, V., and Do Quang, Z. 2009. Bioaerosols in composting facilities: Occupational health risk assessment. *Water Environ. Res.* 81:866–77.

Sigsgaard, T., Bach, B., and Malmros, P. 1990. Respiratory impairment among workers in a garbage handling plant. *Am. J. Indus. Med.* 17:92–93.

Solomon, W. R. 1975. Assessing fungus prevalence in domestic interiors. *J. Allerg. Clin. Immunol.* 56:235–42.

Solomon, W. R. 1976. A volumetric study of winter fungus prevalence in the air of midwestern homes. *J. Allerg. Clin. Immunol.* 57:46–55.

Summerbell, R. C., Krajden, S., and J. Kane. 1989. Potted plants in hospitals as reservoirs of pathogenic fungi. *Mycopathologia* 106:13–22.

Swan, J. R. M., Crook, B., and Gilbert, E. J. 2002. Microbial emissions from composting sites. *Environ. Sci. Technol.* 18:73–101.

Swan, J. R. M., Kelsey, A., Crook, B., and Gilbert, E. J. 2003. *Occupational and environmental exposure to bioaerosols from composts and potential health effects—A critical review of published data.* Rep. 130. Norwich, UK: Health and Safety Executive.

Tekaia, F., and Latgé, J. P. 2005. Aspergillus fumigatus: Saprophyte or pathogen? *Curr. Opin. Microbiol.* 8:385–92.

Varese, G. C., Origione, V., Anastasi, A., Casieri, L., Voyron, S., and Marchisio, V. F. 2002. Airborne fungi in composting plants: A neglected environmental and health hazard. Available at www.biocycle.net.

Venes, D., ed. 1997. *Tabor's cyclopedic medical dictionary.* Philadelphia, PA: F.A. Davis Co.

Viviani, M. A., Tortorano, A. M., and Ajello, L. 2003. Cryptococcus. In *Clinical mycology*, ed. E. J. Anaissie, M. R. McGinnis, and M. A. Pfaller. New York: Churchill Livingston.

Walsh, T. J., Groll, A., Hiemenz, J., Fleming, R., Roilides, E., and Anaissie, E. 2004. Infections due to emerging and uncommon medically important fungal pathogens. *Clin. Microbiol. Infect.* 10(s1):48–76.

Williams, D. L. 1994. (1-3) β-D-glucans. In *Organic dusts*, ed. R. Rylander and R. R. Jacobs. Boca Raton, FL: Lewis Publishers.

Wintermeyer, S. F., Kuschner, W. G., Wong, H., D'Alessandro, A., and Blanc, P. D. 1997. Pulmonary responses after wood chip mulch exposure. *J. Occup. Environ. Med.* 39(4): 308–314.

Wouters, I. M., Spaan, S., Douwes, J., Doekes, G., and Heederik, S. 2006. Overview of personal occupational exposure levels to inhaled dust, endotoxin, β-1(→3)-glucan, and fungal extracellular polysaccharides in the waste management chain. *Annu. Occup. Hygiene* 50(1):39–53.

14 Site Management

INTRODUCTION

In the earlier chapters, the topics of odors, pathogens, and bioaerosols, which are a very important part of site management, were covered. This chapter covers other important aspects of site management. Some, such as runoff and leachate, may be regulated by state or local ordinances. Others, such as nuisance control and housekeeping, are important for public perception, workers, and public health. Health and safety are important for both the public and workers. This could involve pathogen and bioaerosol dispersion, as discussed earlier. A very important part of site management is dust control, noise reduction, vector attraction control, fires, and water management. Each of these topics will be covered in this chapter. This chapter will also discuss sampling, testing, and analysis.

HEALTH AND SAFETY

The health and safety of employees should be a paramount consideration by site managers. Regulations are governed by state and federal government occupational safety and health agencies (OSHA). An important aspect of health and safety is training of employees. Information and instruction should be provided to and discussed with employees.

Health

Exposure to emissions of dust, bioaerosols, and volatile organic compounds is greatest for employees. They not only are exposed to higher concentrations for a given period of time, but also are exposed for longer periods. Table 14.1 summarizes the potential hazardous substances at composting facilities.

Preventive measures at composting sites include personal protective equipment such as dust masks, respirators, and protective clothing. Enclosed buildings should have adequate ventilation. Mobile equipment containing cabs should have filters, and air conditioning should be available where appropriate.

Safety

The safety of employees involves multiple facets. These include personal protection with hard hats and gloves and protection near machinery, especially screens, grinders, shredders, and conveyors. In 1999 at an Ohio composting facility a forty-year-old male machinery operator died of injuries sustained during a fall from a

TABLE 14.1
Potential Hazardous Substances That May Be Found at Composting Facilities

Substance	Item	Information
	Biological	
Bioaerosols	Fungi, bacteria, endotoxin	Can cause allergy, irritation, respiratory infections, inflammation
Dust	Endotoxin, particulates	Irritation, respiratory infections, inflammation
Infective agents	Pathogens—bacteria, viruses, parasites	Diseases
	Chemicals	
Volatile organic compounds (VOCs)	Various organic compounds, e.g., benzene, aliphatic hydrocarbons, acetone	Toxic—irritation, inflammation, illnesses
Nitrogen compounds	Ammonia	Odorous and toxic
Sulfur compounds	Hydrogen sulfide, disulfides	Odorous and toxic

conveyor belt. The National Institute for Occupational Safety and Health (NIOSH) recommended the following:

- Employers should, with the participation of all workers, conduct a job hazard analysis of all worker activities.
- Conveyor lines should be guarded from all belts, conveyors, hoppers, and chute openings. The design of hoppers and chute openings should be such that employees are protected from falling into them.
- Belt conveyors that are fixed in place should have safety guards.
- Employees should be instructed on the rules regarding safety around guards.
- All workers should be thoroughly trained in safety aspects of dangerous equipment, and made aware of all shut-off switches and safety devices.

Where appropriate, guardrails need to be maintained and cleaned. Fire protection equipment needs to be provided and located in areas most likely to be affected. Smoking should be restricted in hazardous and inflammable areas. Employees should be instructed to extinguish all cigarettes before discarding. Eyewash stations and a safety shower should be provided as appropriate.

The following personal protective devices should be available to all employees:

- Hard hats should be worn when instructed or regulated.
- Dust masks and appropriate respiratory devices should be used in areas where employees are exposed to dust and other emissions.
- Earplugs should be worn near noisy machinery.
- Safety glasses should be worn.

SITE MANAGEMENT

Nuisance control involves dust control, noise reduction, and vector control. These topics are rarely discussed and often are given very little attention. However, the public can be very concerned, and it can become an issue. Dust and vector control are also worker and public health concerns.

Dust Control

The primary dust sources are roadways, both paved and unpaved, and screening or grinding operations. Typically, dust is not an issue during windrow turning or moving of composting piles, since the material should be at a moisture content exceeding 45%. Spillage on roads of compost and other materials will dry out and result in dust. Spillage, when possible, should be picked up. Spraying of water, especially on unpaved roads, will reduce dust.

Dust could be generated when the moisture content of the compost is 35% moisture or lower. When compost is screened at this moisture content, dust occurs and is dispersed. Screening should occur at moisture contents between 40 and 45%. Grinding of dry yard waste can also result in dust (Figure 14.1).

Dust during grinding and screening can be reduced by having a mist sprayer over the equipment. The mist should be very fine so as not to increase the moisture

FIGURE 14.1 Dust dispersion during grinding.

content of the material being handled. When a screen is in a building, the use of a dust hood is effective (Figure 14.2).

Dust within a facility is a worker health problem. It can occur on tipping floors, composting floors, and other material handling areas, as shown in Figure 14.3. Dust irritates the eyes and settles in the lungs. It can also result in skin irritation. In one municipal solid waste (MSW) and biosolids composting facility, a worker on a front-

FIGURE 14.2 Dust hood over screen in a composting facility.

FIGURE 14.3 Dust on a tipping floor.

end loader complained of lung problems. Although some of the problems were due to his smoking, a significant problem was due to excessive dust. The front-end loader (FEL) cab was not air-conditioned. Dust contains bioaerosols, especially endotoxins. This may be the reason the literature from Europe, where MSW and biowaste are composted indoors, has indicated greater worker health problems. In enclosed composting facilities, the air ventilation, i.e., air exchanges per unit time, are very important. In the Davenport, Iowa, biosolids and yard waste facility, E&A Environmental Consultants, Inc.'s design specified ten air exchanges per hour during working hours and three when workers were not present. This saved energy.

Dust can settle on equipment and result in fires or corrosion. In one facility, dust in the screen motor resulted in a fire that destroyed the screen. Using a vacuum can be useful to remove dust on motors, conveyor parts, and other areas of the screen. Vacuuming or hosing down a floor at the end of the workday will eliminate dust problems.

Workers should be provided with personal protection devices when working near dusty areas such as screens and tipping floors. These should at a minimum be dust masks or dust filtering respirators.

Noise Control

Noise is a measure of sound pressure and is expressed as decibels (dB). The frequency of a sound is the number of pressure waves the sound generates in a second. Sound levels decrease with distance from the source. Noise is unpleasant and can affect compost facility employees, as well as be a nuisance to nearby neighbors. The perception of noise depends on many factors, such as:

- Loudness
- Frequency
- Location of source
- Location of receptor
- Buffering by vegetation, trees, or a berm
- Individual perception

The source of noise at composting facilities depends on the equipment used, as well as the system. Major noise sources in enclosed facilities are motors, fans, separation equipment, and high-speed equipment. On-site vehicles, trucks, front-end loaders, grinders, and other equipment generate noise as well.

The federal government's Occupational Health and Safety Administration (OSHA) and states regulate noise. In addition, local ordinances may limit noise. It is important to adhere to these regulations to avoid complaints by nearby residences or other potential receptors.

Noise reduction and its impact can be achieved by:

- Using separate enclosures for high-noise-generating equipment
- Requiring employees to wear ear plugs or other noise-canceling equipment
- Buffering the area around the facility
- Constructing berms around the site perimeter

Vector Control

Vectors are animals, insects, or other organisms that carry pathogens. Rats and flies are attracted to fresh compostable waste such as food waste and MSW. Vectors in biosolids composting are regulated by the USEPA 503 regulations. Good housekeeping is paramount to vector control.

An MSW facility, especially the tipping floor, is a source of vector attraction. The area where the waste is brought into the facility should be enclosed and cleaned frequently. Minimal storage of waste should be maintained.

The composting area should be clean and fresh waste removed or incorporated into the composting operation. Standing water should be removed, as it is a source of mosquitoes. This can be done by site grading or treating low spots holding water with lime. Another option is to sop up the water using compost. Roads and grounds should be cleaned regularly. A good-appearing site impresses visitors and neighbors.

Fires

In the past the discussion on fires at composting facilities has been limited—possibly because fires were not frequent or not reported frequently. Much of the information in this discussion was obtained from the following:

Rynk, R., Fires at Composting Facilities: Causes and Conditions, *BioCycle*, 41, 54–58, 2000.

Buggeln, R., and Rynk, R., Self-Heating in Yard Trimmings: Conditions Leading to Spontaneous Combustion, *Compost Sci. Util.*, 10, 162–82, 2002.

Fires have caused extensive damage to some facilities. The worse case was the fire in the Hartford, Connecticut, facility, which was estimated at $27 million in November 1999. It was believed to be caused by spontaneous combustion. This may have resulted from drying of materials due to extended storage. The facility, which used the American Bio Tech system, had twenty fiberglass composting cells, with the composting feedstocks being aerated by forty-eight air lances, each 4.9 m (16 feet) long. At the time of the fire, sixteen vessels contained an estimated 22,940 m^2 (30,000 CY) of material. The design of the facility resulted in compaction and lack of oxygen.

In the Baltimore, Maryland, composting facility, a fire occurred in a storage pile due to carelessness with cigarettes. Rynk (2000) reported that fires in Texas and Georgia were blamed on lightning. At a Philadelphia, Pennsylvania, composting facility sparks from a welder ignited wood chips and stored oil. Welding was also a cause of a fire at the Hamilton, Ohio, facility. At another Ohio composting facility, the finished compost was piled very high using a bulldozer. Dust on a screen motor resulted in a fire at an Iowa facility.

More recently, a fire occurred at a $13 million enclosed facility in Devonshire, England, as a result of a grinder loaded with fuel. Structural damage to the building and roof was considerable.

There have been few recorded fires during aerated static pile or windrow composting operations. More fires have occurred during storage.

Fires can occur at composting facilities for several reasons:

- Spontaneous combustion
- Carelessness from cigarettes
- Overheating of motors, dust on motors
- Sparks from welding operations close to dry material
- Lightning
- Wildfires

Spontaneous combustion is defined as combustion of material in the absence of forced ignition, i.e., without the application of a spark or flame (Buggeln and Rynk, 2002). Spontaneous combustion occurs when materials self-heat to a temperature high enough to ignite the mass.

Fire is chemical oxidation. During the chemical oxidation, heat and light are released as energy (Rynk, 2000). During composting temperature rises because of biological activity. The heat that is generated increases the temperature and, if contained, can exceed 70°C (158°F). Beyond 80°C (176°F) biological activity ceases. If the temperature is allowed to increase and reaches 150°C (302°F) and higher, the organic matter, if dry, could ignite because of chemical reactions. Fire requires fuel (dry organic matter) and oxygen. Composted organic matter contains chemical energy, which can be a source of fuel and support a fire. The amount of energy needed to reach the ignition temperature is called the activation energy. For composting materials, the temperature is in the range of 150 to 200°C (300 to 400°F) (Rynk, 2000). When oxygen is limited and the ignition temperature has been reached, a smoldering fire can occur. Once the pile is opened up and oxygen is available, the fire can develop into a flaming fire.

The causes for spontaneous combustion can be several:

- Very dry material
- Biological activity
- Large, well-insulated piles
- Lack of aeration
- Very high piles
- Compacted piles

There are numerous opportunities to reduce and avoid fires at composting facilities. These are:

- Avoid pile heights in excess of 3.5 m (12 feet).
- Do not compact piles, especially screened and fine material.
- Prevent fires by removing dust around motors. Periodically vacuum motors.
- Instruct operators to avoid discharging cigarettes on the site, and especially into dry piles of materials.

- Be very careful when opening a smoldering pile. It is best to have a front-end loader on hand to open up the pile, and have water hoses and a supply of water.
- Make sure your fire prevention system is adequate and be prepared to address a fire. This includes communication with the local fire department.
- Monitor the piles for hot spots (Naylor, 2004). An excellent and inexpensive way to monitor the temperature of a compost pile is to use a homemade thermocouple. This is accomplished by buying insulated thermocouple wire, removing 1.9 cm (0.75 inch) of insulation, making eight twists, and soldering the ends. Type K (chromel-alumel) or J (iron-constantan) is excellent and inexpensive. Using a simple handheld readout machine ($20 to $30) can provide the temperature or convert the readout to temperature. One can also purchase a multipoint recorder. You can leave these thermocouples in the pile for long periods.
- Store compost at not less than 40% moisture.

Managing Site Water

Site water management predominantly involves leachate and runoff. Leachate is water that percolates through feedstocks or composting material. As precipitation or other sources of water percolate through a pile of compost or feedstock, it will pick up contaminants. These could be pathogens, undesirable soluble inorganic compounds, and organic compounds.

The types of runoff and leachate are:

- Site storm water runoff
- Compost pad storm water runoff
- Compost process leachate
- Compost product leachate

There are several control methods to treat runoff and leachate. Surface water should be directed away from the compost facility, and especially the compost pad. Conditions, as shown in Figure 14.4, can attract vectors transmitting diseases and be a safety hazard.

It is important to manage the process to minimize runoff and leachate. Leachate needs to be avoided, but if it occurs, it should be collected and treated. Collected leachate can be used either in the process to provide additional moisture, providing it is not used toward the end of the process, or on the finished product. It may be added into the composting material before initiating regulatory compliance. If possible, it can be discharged to a municipal wastewater facility or to an on-site wastewater treatment. Using covers, as described earlier, will also minimize leachate generation. Housekeeping is very important.

Leachate and runoff should be avoided from entering watercourses. This can be accomplished by having a grassy area to collect the leachate or runoff prior to entering the watercourse. Runoff/leachate could be diverted to a holding basin and used either on the site during dry periods, for dust control on unpaved roads, or for irrigating adjacent land. The water could be available for fire prevention. Leachate should

FIGURE 14.4 Runoff and leachate on a composting pad.

be prevented from entering groundwater, especially where the groundwater table is high. This may require paving of areas most prone to produce leachate. In some areas where paving is expensive, soil cement can be used. These considerations may require local or state approval.

The design of the facility should minimize or prevent runoff and standing water. Standing water from either leachate or precipitation should be removed. It can attract flies and mosquitoes, which are vectors. Furthermore, it can be a source of site odors. The site should have a slope to drain the runoff/leachate. Compost can be used to fill and sop up water in puddles. Lime can also be used to destroy mosquito larvae.

SAMPLING, ANALYZING, AND TESTING

Sampling, analyzing, and testing the compost product are very important steps for several reasons:

- Meeting state and federal regulations
- Reporting to authorities
- Making sure your product does not contain any undesirable contaminants that could be harmful to plants, animals, and humans
- Good public relations
- Knowing your product for good marketing

Compost facilities must comply with federal, state, and local rules and regulations.

The U.S. Environmental Protection Agency regulates biosolids composting for both pathogens and several inorganic elements. Organic elements are not regulated. The regulations involve:

- Heavy metals (trace elements)
- Temperature as related to time
- Pathogens and vector attraction

Most states incorporate these requirements for other feedstocks.

The U.S. Composting Council (USCC) provides information under the Test Methods for the Examination of Composting and Compost (TMECC), which is a laboratory manual modeled after the American Society for Testing and Materials (ASTM):

The first edition is a collection of all known methods.

The second edition is currently under way, and will include new methods and amendments to the old. This is a very comprehensive document and will not be repeated here.

Essentially, there are seven chapters detailing sampling, collection, and laboratory preparation; physical parameters; chemical properties; organic and biological properties; synthetic organic compounds; and pathogen testing.

Excluding the need to meet regulations, why do we test compost?

- Health and safety
- To know your product
- Information on physical attributes
- Knowledge of the nutrient constituents
- To provide information on application rates
- To provide information on end use suitability
- To provide information on the safety of your product
- To solve problems related to the process

Several laboratories follow the TMECC procedures. Most of the laboratories report the following:

- Nutrients
- Trace metals and heavy metals
- Salts
- pH
- Bulk density
- Bacteria
  - Salmonella
  - Fecal coliform
- Respiration
- Maturity

These parameters are reported in their values, e.g., percentage, mg/kg, or units.

Sampling needs to be done in a specific manner with careful recording and labeling of the samples. For example, when sampling for pathogens, it is important to make sure all sampling tools and collection materials are sterile. When sampling and sending samples to the laboratory, follow the lab's directions. Here are some suggested procedures:

- All containers for sampling should be sterile. This can be done by using new containers that have been sterilized and are available from laboratories or some pharmacies.
- The tools for sampling should be sterilized. Take a shovel, pour alcohol on it, and ignite it. Use the shovel when cold to open pile. Take a metal trowel with an insulated handle. Pour alcohol and ignite. When cool, use it to collect samples from various areas in the pile.
- Place the samples in the sterile containers and close tightly.
- Label all containers with the date, time of sampling, and location, i.e., compost pile, windrow, curing, or storage pile.
- Send to laboratory.
- Do not use galvanized pails or other metal containers if you are sampling for heavy metals. Use plastic containers.

Testing and analysis need to be done correctly and by knowledgeable laboratories. They need to follow the TMECC procedures, and these need to be identified in the laboratory report.

It is important that the marketer, whether a specific individual or the facility operator, knows what these values mean and their importance to the user.

How can you use the testing information?

- Provide regulators with information that you are meeting regulations. This could avoid problems and expenses. As an example, one facility did not keep good records of its temperature data showing that they met the federal regulations. As a result, the USEPA regional office did not allow it to market, dispose, or use its product unless additional costly tests for parasites and viruses were done.
- Adjust your operation to provide the highest-quality product.
- Inform the public of the safety of your product.
- Inform the public of the value of your product.
- Inform the public of how best to use the product.
- Provide your marketer with information.

CONCLUSION

The subjects under the site management section deal with nuisances such as dust, noise, vectors, fires, and excess site water consisting of leachate and runoff. All of these need to be considered during design and construction of a facility and need to be prevented during operations. Dust, noise, and fires can illicit public complaints.

Vectors are a health problem, which could affect both the public and workers. Leachate and runoff could result in odors and harboring of flies and mosquitoes.

The section on sampling, testing, and analysis refers primarily to the U.S. Composting Council Test Methods for the Examination of Composting and Compost (TMECC). This information should be available at each facility. Analytical information during operations can suggest that modifications may be needed to produce a better product. Some of the information obtained is essential for compliance with regulatory requirements. The analytical information is very useful to good marketing. It informs the user of what the product contains and how best to use it. It assures the user that hazardous contaminants are not in the product.

REFERENCES

Buggeln, R., and Rynk, R. 2002. Self-heating in yard trimmings: Conditions leading to spontaneous combustion. *Compost Sci. Util.* 10:162–82.

Naylor, L. 2004. Fire prevention at composting, mulch facilities. *BioCycle* 45:30–32.

Rynk, R. 2000. Fires at composting facilities: Causes and conditions. *BioCycle* 41(1):54–58.

15 Public Relations, Communication, and Regulations

INTRODUCTION

PUBLIC RELATIONS AND COMMUNICATION

Public relations are a very important component of the operation of a facility. It is important to begin communication with the public as early as possible. It is necessary for the community to be aware of the risk, such as the potential for odors; the benefits to the environment; possible savings in collection costs; and any remuneration to the community, if applicable. Therefore, education, publicity, and communication are very important aspects of public relations.

It is important to communicate to the public what are the alternatives to composting and what these alternatives will cost. Today's public is much more sophisticated and is willing to support technologies that help the environment. For example, how many people know that the addition of compost will sequester carbon for the mitigation of global warming? In comparison, landfilling and incineration (other solid waste disposal alternatives) add pollutants toward global warming. Compost adds carbon to the soil. Aerobic composting does not emit methane, a greenhouse gas that a landfill does.

Public relations is susceptible to human subjectivity. People are influenced by other people and often accept what they say regardless if it is true or false. Public relations and communication can counteract misinformation as well as promote truths. This often becomes very evident in public meetings. People become emotional and angry. The people opposing a facility feed on fear, especially as related to health. They will ignore evidence from research or even contradict or negate the words of an expert in the field.

One of the most interesting and successful public relations and communication programs that involves compost and other environmental aspects is Soils for Salmon in the state of Washington. This program was begun in 1999 by the Washington Organic Recycling Council. One aspect of this program is to promote healthy soil. They promote the idea that water quality is very important for saving the salmon in the northwest. In the process of urban expansion and land development, surface runoff increases and water quality decreases. They point out that deterioration of soil structure, because of human activity, can result in high runoff and erosion, poor vegetation, shallow root growth, and limited beneficial organisms. A healthy soil promotes good, vigorous vegetation; retains water; increases infiltration, thus reducing runoff and erosion; and binds and breaks down pollutants. Their literature states "compost and other organics can improve soil health and environmental functions." Runoff and erosion contaminate streams and destroy the ecosystem.

This program has received both regional and national recognition as a practical approach to link the benefits of healthy functioning soils with clean, healthy water resources, and thus benefit the salmon.

A good public relations program and good communication with the public regarding the composting facility, its function, and benefits are very important.

Regulations

The first regulations pertaining to compost were the federal regulations published in the *Federal Register* (58 FR 9248 to 9404) on February 1993, entitled "The Standards for the Use or Disposal of Sewage Sludge" (Title 40 of the Code of Federal Regulations (CFR), Part 503). The 503 rule covered pollutant limits and pathogen and vector attraction. A major part of this regulation was titled "Control of Pathogens and Vector Attraction in Sewage Sludge." This section was revised in 2003. The pollutant limits pertained to trace elements (often referred to as heavy metals). These are arsenic (As), cadmium (Cd), chromium (Cr), copper (Cu), lead (Pb), mercury (Hg), molybdenum (Mo), nickel (Ni), selenium (Se), and zinc (Zn). Subsequently, Cr was deleted from the regulations. The federal regulations did not regulate organic compound contaminants. They did evaluate fourteen organic compounds.

States adopted these regulations and applied them to other wastes or waste products, such as compost, in addition to sewage sludge and biosolids. Besides regulations on pollutants and pathogens, states have odor regulations:

- There are no federal regulations regarding odors in the United States.
- Odor nuisance regulations are in forty-two states.
- Olfactometry-based regulations (dilution-to-threshold (D/T)) are used in eight states.
- Some cities, regions, or districts have their own olfactometry regulations.

There are few state, regional, or local requirements as to the physical properties of the compost. The Canadian government and provinces have regulations on physical properties of compost.

The most important aspect of the federal and state regulations with respect to compost disinfection is strict adherence to the time-temperature requirements. If these are not met, considerable expenses can result for sampling of viruses and parasites.

PUBLIC RELATIONS, PARTICIPATION, AND COMMUNICATION

Beecher et al. (2004) stated it is not public relations; it is public relationship. From the beginning, develop trust. The public should be involved from the beginning during the earliest part of the planning process. Public agencies generally set up a citizen's advisory committee. The committee should be all-inclusive and not only include neighbors or concerned citizens, but also potential users, university personnel, and public officials. This would give a more balanced voice during discussions. Get input from stakeholders. It is also important to begin the educational program early. This could consist of lectures to local organizations such as the Lion's Club, League of Women's Voters, etc. Brochures or fliers should be disseminated. A periodical newsletter should be prepared. Some organizations provide an 800 telephone number and designate a person on the staff who will answer and provide information or transmit the complaints. This could be useful in identifying the source of odors and conditions that prevail at the time. It may then be possible to modify procedures to minimize the situation. Also, informing the complaining person that you are actively doing something reduces the concern or tension.

Public meetings are often necessary. They are not always the best place for communication, especially if there are angry, distrustful persons who feel threatened (Beecher et al., 2004).

Beecher et al. (2004) recommended the following planning and advance work for a public hearing:

- Talk to community leaders and representatives in advance.
- Determine goals for the meeting.
- Arrange the venue.
- Do not have an "expert" stand up in front and talk on and on.
- Ensure opportunities for all perspectives to be shared.
- Visibly record concerns.
- Answer questions respectfully.
- Listen and hear.
- Have an independent monitor.

In addition, this authors experience has been as follows:

- Use local experts who have respect and credibility in the community. This author has been to hearings where a worldwide expert from afar was brought in. Regardless of his or her standing, the community was not impressed. It would have been more desirable to bring someone from the local university who is trusted.
- To avoid mayhem, request initially that written questions be submitted and identify persons who wish to speak at the beginning of the meeting, and restrict them to a short presentation.

If you cannot answer the question, identify a staff member with his or her background to answer the question. If this is not possible, inform the individual that you will try to get the answer and do so. Obtain an answer from a reliable source, such as an agricultural university, and provide the answer to the inquirer. Do not put down a person. Beecher et al. (2004) stress that trust is extremely important. If there is no trust by the public, there is suspicion that "the wool is being pulled over the eyes." The staff needs to be involved and honest with visitors or callers.

Implementing demonstration projects, attendance, and exhibits at flower or horticultural shows help promote the project and its value. It is important to provide tours to the public and schools. Often children attending these projects are very involved in recycling and environmental issues and can impress upon their parents the value of these projects. At times provision of compost to neighbors for promoting gardens and enhanced landscaping has proved to be useful. Assistance in establishing parks or playgrounds could also be useful.

REGULATIONS

The federal regulations adopted by states relate to pollution (trace elements or heavy metals), pathogens, and vector attraction. States, regions, or local entities can make the regulations stricter. They cannot make them less strict.

Compost federal regulations pertaining to sewage sludge, septage, and biosolids fall under 40CFR503, which was promulgated in 1993 under the authority of the Clean Water Act, as amended in 1997, and the 1976 Resource Conservation and Recovery Act. The 503 regulations replaced 40CFR257. Biosolids or biosolids products can be land applied in bulk, sold, or given away in bags or other containers (buckets, boxes, or cartons, or vehicles with a load capacity of less than 1 metric ton).

Concepts and Approaches to Regulations

Kennedy (1992) presented three basic approaches to the development of regulations as related to product use:

- No-net degradation
- Risk-based approach
- Best achievable approach

The no-net degradation concept is based on the premise that the application of compost should not increase the level of a heavy metal or other contaminant in the soil. Several European countries and Canadian provinces have set guidelines or regulations based on this concept. One problem with no-net degradation is the question of what should be used as a soil base level. Soil quality varies greatly within a small area. Urban soils may have higher levels of lead from leaded gasoline than rural areas. Rural land that has had manure applied would have different concentrations of pollutants than land that has not had manure applied. Regional standards will have to be established based on fluctuations in soil quality. If the no-net degradation concept was used on a site-by-site basis, it would create excessive sampling requirements and would allow the use of lower-quality material on areas that are already contaminated. Another problem with the no-net degradation concept is that soils are continuously amended with fertilizers, pesticides, herbicides, and other chemicals. This not only changes the baseline quality of the soil, but also illustrates the illogic in singling out a single material as the only regulated material.

The best achievable approach ignores health and environmental aspects and primarily considers technology and economics. Standards are based on what technology can achieve.

The risk-based approach considers the potential risk to humans, animals, plants, and soil biota, as well as environmental consequences. This approach evaluates the potential toxic effects of a chemical on the individual (human, animal, or plant) or environmental entity. The risk-based approach considers the risk in relation to other risks in the environment. This approach is dependent on having sufficient good data. The most comprehensive risk evaluation performed was for heavy metals by the USEPA 40CFR503 regulations for the disposal and use of biosolids. This approach was not used for pathogens. Table 15.1 compares the risk-based approach to the no-net degradation approach for Switzerland and Ontario, Canada. The concentrations of heavy metals in the no-net degradation approach are much lower.

U.S. federal regulations dealing with land application of biosolids or biosolids products are under the jurisdiction of the Environmental Protection Agency (USEPA). Enforcement is through USEPA regions with the aid of state regulatory agencies. Those states that have delegation have regulatory responsibility. Regulations promulgated by USEPA cover biosolids or any material containing biosolids. These regulations were required by the Clean Water Act Amendments of 1987 (Sections 405(d) and (e)) as amended (33 USCA 1251, et seq.). The regulations were published in the *Federal Register* (58 FR 9248 to 9404) as "The Standards for the Use or Disposal of Sewage Sludge," Title 40 of the Code of Federal Regulations, Part 503. The 503 rule was published on February 19, 1993, and became effective on March 22, 1993. The 503 rule was amended on February 25, 1994 (59 FR 9095) for molybdenum. The

TABLE 15.1
Comparison between Risked-Based and No-Net Degradation Standards for Heavy Metals in Compost

Heavy Metal Element	Risked Based mg/kg Dry Matter	Proposed EU mg/kg Dry Matter	No-Net Degradation ppm	No-Net Degradation ppm
	United States—EPA		Switzerland	Ontario, Canada
Cadmium	39	10	3	3
Chromium	NR	1,000	150	50
Copper	1,500	1,000	150	50
Mercury	17	10	3	0.15
Nickel	420	300	100	60
Lead	300	750	150	150
Zinc	2,800	2,500	500	500

Source: Based on Harrison, E. Z. and Richard, T. R. *Municipal Solid Waste Composting: Issues in Policy and Regulation*, Fact Sheet 6, Cornell Waste Management Institute, Ithaca, NY, 2005; Siebert, S. in *End-Waste Workshop on Compost, 22nd March 2007*, Bundesgutemeinschaft, Kompost e.v., Sevilla, Spain, 2007.

pollutant concentration limits and annual pollutant loading rates for molybdenum were deleted. Only the ceiling concentration limit of 75 mg/kg was retained.

Two other pollutant (Cr and Se) limits were contested in the courts. The Leather Industries of America, Association of Metropolitan Sewerage Agencies, Milwaukee Metropolitan Sewerage District, and City of Pueblo, Colorado, filed lawsuits. On March 5, 1993, the Leather Industries of America, Inc. filed a petition with the U.S. Circuit Court of Appeals seeking review of the pollutant limits for Cr. On June 17, 1993, the City of Pueblo, Colorado, filed a petition for review with the U.S. Court of Appeals challenging the Se pollutant limits. Soil and water in that area contain high levels of Se. On October 25, 1995, the USEPA deleted the pollutant limits for Cr and modified the Se limit to 100 mg/kg.

These actions point out to the very important and significant aspect of regulations vs. guidelines. Regulations can be changed. It also shows how the U.S. regulations can be modified if new data become available or the regulations are not equally applied to all aspects. In addition to heavy metals, the 503 rule regulates pathogens and vector attraction. On December 23, 1999, the USEPA published in the *Federal Register* (Volume 64, Number 246, pages 72045–72062) a proposal to amend the management standards for sewage sludge. A numeric concentration limit is proposed for dioxin and dioxin-like compounds in sewage sludge that is applied to the land, as well as monitoring, record keeping, and reporting requirements for dioxins in sewage sludge that is land applied.

Much of the discussion in this section is from four USEPA documents.

1. *Federal Register*, Standards for the Use or Disposal of Sewage Sludge, Final Rules, Part II, 40 CFR Part 257 et al., Environmental Protection Agency, February 19, 1993
2. U.S. Environmental Protection Agency, Office of Wastewater Management (4204), *A Plain English Guide to the EPA Part 503 Biosolids Rule*, EPA/832/R-93/003, September 1994
3. U.S. Environmental Protection Agency, Office of Wastewater Management (4204), Guide to the Biosolids Risk Assessments for the Part 503 Rule, EPA832-B-95-005, unpublished document, courtesy of Dr. J. Walker
4. U.S. Environmental Protection Agency, Office of Research and Development, Environmental Regulations and Technology, *Control of Pathogens and Vector Attraction in Sewage Sludge*, EPA/625/R-92/013, revised October 1999 and 2003

The 503 rule was designed to protect public health and the environment from "any reasonably anticipated adverse effects of certain pollutants and contaminants that may be present in [biosolids]" (USEPA, 1994). The USEPA clearly stated that it promotes the beneficial use of biosolids. A very intensive risk assessment was conducted. The rule making took nine years to complete and evaluated research from the past twenty-five years. In 1984, the USEPA considered two hundred pollutants identified in the 40 Cities Study. The selection of the two hundred pollutants was based on the following criteria:

- Human exposure and health effects
- Plant uptake of pollutants
- Phytotoxicity
- Effects in domestic animals and wildlife
- Effects in aquatic organisms
- Frequency of pollutant occurrence in biosolids

This list of pollutants was submitted for review by four panels. The panels recommended that approximately fifty of the two hundred pollutants listed be further studied. In the final regulations, the USEPA addressed twenty-five pollutants using fourteen exposure pathways (Ryan and Chaney, 1995). The twenty-five pollutants were:

Organics	Heavy Metals
Aldrin/dieldrin (total)	Arsenic
Benzene	Cadmium
Benzo(a)pyrene Chromium	Copper
Bis(2-ethylhexyl)phthalate	Lead
Chlordane	Mercury
DDT/DDE/DDD (total)	Molybdenum
Heptachlor	Nickel
Hexachlorobenzene	Selenium
Hexachlorobutadiene	Zinc
Lindane	
N-Nitrosodimethylamine	
Polychlorinated biphenyls	
Toxaphene	
Trichloroethylene	

There were four basic steps to the risk assessment (USEPA, 1995):

- Hazard identification: Can the identified pollutants harm human health or the environment?
- Exposure assessment: Who is exposed? How do they become exposed? How much exposure occurs? Highly exposed individuals were identified, and their exposure to pollutants in biosolids evaluated. Fourteen exposure pathways were identified for land application of biosolids (Table 11.1).
- Dose-response evaluation: The likelihood of an individual developing a particular disease as the dose and exposure increase. The two EPA toxicity factors used whenever available were (1) risk reference doses (RfDs)—daily intake, and (2) cancer potency values (q_1*s)—conservative indication of the likelihood of a chemical inducing or causing cancer during the lifetime of a continuously exposed individual.
- Risk characterization: What is the likelihood of an adverse effect in the population exposed to a pollutant under the conditions studied? Risk is calculated as

$$\text{Risk} = \text{Hazard} \times \text{Exposure}$$

Hazard refers to the toxicity of a substance determined during the hazard identification and dose-response evaluation; exposure is determined through the exposure assessment (USEPA, 1995). The USEPA made a policy decision to regulate risk at 1 × 10–4. Ryan and Chaney (1995) discuss the general USEPA approach.

Heavy Metal Regulations

Table 15.2 shows the USEPA exceptional quality (EQ) concentration limits for heavy metals and some other elements regulated in biosolids products. The term *exceptional quality* was designed for use of waste products on lawns, home gardens, or that were either sold or given away.

TABLE 15.2
Pollutant Limits for Heavy Metals in Biosolids and Biosolids' Products

Pollutant	Ceiling Concentration Limits for All Biosolids Applied to Land mg/kg[a]	Pollutant Concentration Limits for EQ and PC Biosolids mg/kg[a]	Cumulative Pollutant Loading Rate Limits for CPLR Biosolids kg/ha	Annual Pollutant Loading Rate Limits for APLR Biosolids kg/ha/365-day period
Arsenic	75	41	41	2.0
Cadmium	85	39	39	1.9
Copper	4,300	1,500	1,500	75
Lead	840	300	300	15
Mercury	57	17	17	0.85
Molybdenum[b]	75	—	—	—
Nickel	420	420	420	21
Selenium[d]	100	36	100	5.0
Zinc	7,500	2,800	2,800	140
Applies to:	All biosolids that are land applied	Bulk biosolids and bagged biosolids[c]	Bulk biosolids	Bagged biosolids[c]
From Part 503	Table 1, Section 503.13	Table 3, Section 503.13	Table 2, Section 503.13	Table 4, Section 503.13

Source: USEPA, 1995.

[a] Dry-weight basis.

[b] The limits for molybdenum were deleted from the 503 rule on February 25, 1994 (*Federal Register*, Vol. 39, No. 38, p. 9095).

[c] Bagged biosolids sold or given away in a bag or other container.

[d] Chromium deleted from regulations and selenium modified in 1995.

Pathogen and Vector Attraction Rules

In addition to pollutant limits, the 503 rule also requires pathogen and vector attraction reduction criteria. The basis for the 503 pathogen requirements are provided in the USEPA document *Technical Support Document for Reduction of Pathogens and Vector Attraction in Sewage Sludge* (USEPA, 1992). In 1999, the USEPA issued a revision of the document *Environmental Regulations and Technology Control of Pathogens and Vector Attraction in Sewage Sludge* (EPA/625/R-92-013, revised October 1999). A further revision occurred in 2003. In the previous USEPA 257 regulations, the only requirements for composting were based on time-temperature relationships. A 1988 study (Yanko, 1988) demonstrated that regrowth of pathogens occurs in biosolids compost. In this study salmonellae were detected 165 times in 365 measurements. No salmonellae were detected in the eighty-six measurements for which the fecal coliform densities were less than 1,000 MPN (most probable number) per gram. This indicated that the potential for finding salmonellae would be highly unlikely when the fecal coliform densities were less than 1,000 MPN per gram. The correlation between fecal coliform densities and frequency of salmonellae detection is shown in Figure 11.1 (USEPA, 1992; Farrell, 1992). The USEPA (1992) states that the reason for alternatively using either the fecal coliform test or the salmonellae test is that fecal coliform can regrow to levels exceeding 1,000 MPN/g, but salmonellae, once totally eliminated, can never grow.

The vector reduction criterion, which applies to composted biosolids that contain partially decomposed organic bulking agents, requires that the biosolids be aerobically treated for fourteen days or longer, during which time the temperature must always be over 40°C and the average temperature must be higher than 45°C. Class B biosolids compost, in addition to heavy metal and pathogen reduction criteria, also has site restrictions. These site restrictions primarily apply to use on food crops, animal grazing, turf growing, and public access (USEPA, 1995).

The 503 regulations also provide sampling and analysis methodologies. One of the most important aspects of the 503 regulations, which affect land application methodologies, is liability. Direct land application, whether by a public or private entity, is the legal responsibility of the producer of biosolids. If a municipality or its contractor violates the permit requirement for land application of biosolids, the producer, its employees, and the contractor are subject to civil and criminal action. For example, if a contractor violates the municipality's permit to apply a specific quantity of biosolids containing the 503 heavy metal limitations, the contractor, the municipality, and any knowledgeable individuals can be liable and sued for both criminal and civil damages.

The distribution and marketing of biosolids products, such as compost, does not entail similar liability. A contractor or individual purchasing compost containing the limit of heavy metals and distributing or marketing the compost at excessive rates does not face criminal or civil charges. Only product liability litigation could result. For example, if compost is provided to a user without adequate instruction on its use and it causes phytotoxicity, the provider may have liability.

Pathogens in sewage sludge, biosolids, and septage are regulated under Subpart D of the Part 503 rule (USEPA, 1999). Two classes are designated, class A and class B.

Class A is designed so that pathogens are not detected in biosolids or biosolids products. These include bulk or bagged products that are given away for home gardens or other horticultural uses. Once the sewage sludge is treated to meet class A or class B, it can be designated as biosolids. This distinguishes it from the untreated material.

The pathogen regulations involve three aspects:

- Specific pathogen requirements
- Process requirements
- Vector attraction requirements (VARs)

1. Class A pathogen requirements are as follows:
   The density of fecal coliform in the sewage sludge or biosolids must be less than 1,000 most probable number (MPN) per gram total solids (dry-weight basis).
   OR
   The density of *Salmonella* sp. bacteria in the sewage sludge or biosolids must be less than 3 MPN per 4 gram of total solids (dry-weight basis).
   Either of these requirements must be met at one of the following times:
   - When sewage sludge or biosolids are used or disposed.
   - Sludge or biosolids products such as compost, alkaline stabilized material, or heat-dried products are prepared for sale or give-away in a bag or other container for land application; or
   - When the sewage sludge or biosolids product or derived materials are prepared to meet the requirements for EQ biosolids.
   - Pathogen reduction must take place before or at the same time as VAR process requirements.
2. Processes to Further Reduce Pathogens (PFRP)
   - Using either the within-vessel composting method or the static aerated pile composting method, the temperature of the sewage sludge is maintained at 55°C (131°F) or higher for three consecutive days.
   - Using the windrow composting method, the temperature of the sewage sludge is maintained at 55°C (131°F) or higher for 15 consecutive days or longer. During the period when the compost is maintained at 55°C (131°F) or higher, there shall be a minimum of five turnings of the windrow.
3. Process to Significantly Reduce Pathogens
   This is generally of importance to composting, except for very small homes or composting toilets. Composting facilities can very easily meet and should meet PFRP regulations.

If a vendor or individual desires to have a specific process accepted by the USEPA, he or she must submit all documentation and data to EPA's Pathogen Equivalency Committee. This can be done for a process to be accepted for a local application or for the process to be accepted on a national level (USEPA, 2003).

ODOR REGULATIONS

There are no federal regulations regarding odors in the United States. States regulate odors using either nuisance regulations or olfactometry-based (dilution-to-threshold (D/T)) regulations. Nuisance regulations are promulgated in forty-two states, and dilution to threshold in six states. Some cities, regions, or districts have their own olfactometry regulations.

There are different odor approaches, guidelines, and regulations in the United States (Mahin, 2001; Mahin et al., 2000). The approaches are:

1. One approach is the use of ambient limits for individual compounds such as hydrogen sulfide. Because of the myriad compounds that cause odors from various sources, this approach raises problems. Table 15.3 provides some data on ambient standards.
2. A second approach is off-site nuisance or annoyance conditions as determined by field inspectors in response to complaints from the public. This is a simple approach but has legal issues. Furthermore, it often occurs that several neighbors get together and complain, although only one has smelled an odor. This is a vague approach.

Table 15.4 provides a summary of odor regulations in some states and authorities. Most states do not have regulations. There is a wide discrepancy among those states or authorities. Some use an odor assessment at the property line, whereas others use the receptors. Regulations of less than 5 to 7 D/T are not sensible. The body odor of the regulator will exceed that amount. Measurement at the property line also does

TABLE 15.3
Ambient Odor Standards Using Limits for Individual Compounds

Location	Compound	Ambient Odor Standard
California	Hydrogen sulfide	30 ppbv* (1 h average)
Connecticut	Hydrogen sulfide	6.3 μg/m³
	Methyl mercaptan	2.2 μg/m³
Idaho	Hydrogen sulfide	10 ppbv (24 h average)
		30 ppbv (30 min average)
Minnesota	Hydrogen sulfide	30 ppbv (30 h average; not to exceed more than 2 days in a 5-day period)
		50 ppbv (50 min average; not to exceed more than 2 times per year)
Nebraska	Total reduced sulfur	100 ppb (30 min average)
New Mexico	Hydrogen sulfide	Hydrogen sulfide 10 ppbv (1 h average) or 30–100 ppbv (30 min average)
New York City	Hydrogen sulfide	1 ppbv for WWTP
North Dakota	Hydrogen sulfide	50 ppbv (instantaneous; two readings 15 min apart)
Pennsylvania	Hydrogen sulfide	100 ppbv (1 h average)
		5 ppbv (24 h average)
Texas	Hydrogen sulfide	80 ppbv (30 min average for residential and commercial and 120 ppbv for industrial, vacant, or range lands)

Source: Data from Mahin, T. D. *Water Sci. Technol.*, 44(9), 87–102, 2000.

TABLE 15.4
Summary of Odor Regulations in Some States and Authorities

State	Odor Regulation
California	No statewide requirement; 5 D/T at the property line has been used for WWTPs
City of San Diego WWTP	5 D/T 5 min intervals
Bay Area Air Quality District	5 D/T if 10 nuisance complaints generated in 90 days
Colorado	7 D/T measured off-site twice within 1 h
Connecticut	7 D/T field measurement by state inspector 3 times in an hour
Florida	No statewide requirement
Palm Beach County Solid Waste Authority	7 D/T used in compost facility design
Illinois	No statewide requirement
Kankakee Wastewater Authority, IL	4 D/T off-site in 2 min intervals
Massachusetts	5 D/T at the property line
New Jersey	5 D/T at receptors
North Carolina	4 D/T off-site in 30 sec intervals
Oregon	1 to 2 D/T 15 min intervals
Pennsylvania	"No detectable malodor"; 4 D/T off-site has been used
Washington	5 D/T at 5 min intervals used for the city of Seattle WWTP

Source: Data from Mahin, T. D. et al., *Water Environ. Technol.*, 12, 49–53, 2000.

not make sense. Receptors may be far away or the operations may be very close to the property line.

Generally, regulations in Europe and several other countries for heavy metals or elements are more restrictive than those in the United States. Pathogen regulations are essentially the same. Table 15.5 provides some information from other countries in comparison to the USEPA regulations for biosolids compost. It is difficult to compare the European standards to those of the USEPA since the latter particularly refers to biosolids products only, whereas the European standards are primarily for biowaste. Levels of inorganic elements in biowaste, which is primarily separated household waste, consisting of food waste and unrecycled material, are much lower.

CONCLUSION

This chapter covered two important subjects: (1) public relations and communication, and (2) regulations. There is a relation between these two subjects. Compost regulations are designed to protect the public. They also provide the public with assurance that the product is safe. These aspects are very important to convey to the public the value of having a composting facility.

Public relations and communications must begin early in the planning process. These actions can avoid permitting delays, as well as delaying the construction of a facility.

TABLE 15.5
Comparison of Heavy Metal Concentrations in Compost for Several Countries

Country	Cr	Ni	Cu	Zn	Cd	Hg	Pb
USEPA biosolids	—	420	1,500	2,500	39	17	250
Canada CCME cat. A	210	62	100	1,850	3	0.8	150
EC draft biowaste class 1	100	50	100	200	0.7	0.5	100
UK Composting Assoc. Qual.	100	50	200	500	1.5	1	150
Belgium	70	20	90	300	1.5	1	120
Germany biowaste ordin. 1	70	35	70	300	1	0.7	100
Netherlands	50	20	60	200	1	0.3	100
Australia limits for biosolids	400	60	200	250	3	1	200

Source: Hogg, D. et al., *Comparison of Compost Standards within the EU, North America and Australia*, Waste and Resources Action Programme (WRAP), The Old Academy, Banbuey, UK, 2002.

The regulations in the United States are divided into two major sections. One section deals with pollutants and, at the present time, refers to heavy metals and other elements. Organic pollutants are not regulated, but may be in the future. The second section covers pathogens and vector attraction. This document has been revised several times. The most recent addition was printed 2003.

REFERENCES

Beecher, N., Connell, B., Epstein, E., Filtz, J., Goldstein, N., and Lono, M. 2004. *Public perception of biosolids recycling: Developing public participation and earning trust*. Alexandria, VA: Water Environment Federation, Water Environment Research Foundation.

Harrison, E. Z., and Richard, T. R. 2005. *Municipal solid waste composting: Issues in policy and regulation*. Fact sheet 6. Ithaca, NY: Cornell Waste Management Institute.

Hogg, D., Barth, J., Favoino, E., Centemero, M., Caimi, V., Aminger, F., Devliegher, W., and Antler, S. 2002. *Comparison of compost standards within the EU, North America, and Australia*. Banbury, UK: Waste and Resources Action Programme (WRAP), The Old Academy.

Kennedy, J. 1992. A review of composting criteria. Proc. of the Composting Council of Canada. Second Annual Meeting, Ottawa, ON, Canada.

Mahin, T. D. 2001. Comparison of different approaches used to regulate odors around the world. *Water Sci. Technol.* 44(9):87–102.

Mahin, T. D., Pope, R., and McGinley, M. 2000. When is a smell a nuisance? *Water Environ. Technol.* 12(5):49–53.

Ryan, J. A., and Chaney, R. L. 1995. Issues of risk assessment and its utility in development of soil standards: The 503 methodology as an example. In *Biogeochemistry of trace elements*. Paris.

Siebert, S. 2007. European quality assurance for compost. In *End-waste workshop on compost, 22nd March 2007*. Sevilla, Spain: Bundesgutemeinschaft, Kompost e.v.

USEPA. 1992. Technical support document for reduction of pathogens and vector attraction in sewage sludge. EPA 822/R-93-004.

USEPA. 1994. A Plain English guide to the EPA Part 503 biosolids rule. U.S. Environ. Protection Agency. EPA/832/R-93/003.

USEPA. 1995. Standards for the Use or Disposal of Sewage Sludge; Final Rule and Proposed Rule, U.S. Environmental Protection Agency, Federal Register Vol. 60, No. 206, Washington, DC.

USEPA. 1999. Control of Pathogens and Vector Attraction in Sewage Sludge., U.S. Environmental Agency, Office of Research and Development, National Risk Management Research Laboratory, Center for Environmental Research Information. EPA/625/R-92-013, Cincinnati, OH.

USEPA. 2003. Control of Pathogens and Vector Attractiobn in Sewage Sludge., U.S. Environmental Protection Agency, Office of Research and Development, National Risk Management Research Laboratory, Center for Environmental Research Information. EPA/625/R-92-013, Revised July 2003. Cincinnati, OH.

Yanko, W. A. 1988. Occurrence of pathogens in distribution and marketing municipal sludges. Rept. No. EPA-6/1-87-014. (NTIS PB 88-154273/AS.) Cincinnati Health Effects Res. Laboratory, Cincinnati, OH.

16 Product Utilization and Marketing

INTRODUCTION

A major aspect of the design, management, and operations of a composting facility is to produce a high-quality, marketable product. Rarely does the sale of compost compensate for the capital and operating cost of a composting facility. Most private facilities rely on a tipping fee or the equivalent. However, the sale of the compost can significantly reduce operating costs. Public facilities can cover some or all of the capital and operational and maintenance (O&M) costs from various sources. These could include sewage fees, municipal solid waste (MSW) collection fees, yard waste tipping fees, and the sale of compost.

Although utilization is not directly a component of compost technology, it is greatly affected by processing and is an important aspect of marketing. Understanding how best to utilize compost and the factors that make a product valuable requires an understanding of the composition of compost and its potential benefits.

Compost is an organic matter resource that predominantly improves soil physical properties. In addition, it provides various levels of macro plant nutrients such as nitrogen (N), phosphorus (P), potassium (K), calcium (Ca), and magnesium (Mg), as well as many of the micronutrients required for plant growth. The level of macro- and micronutrients for plants depends to some extent on the feedstock. Nitrogen is the one element that is reduced during the composting operation. Compost also enhances the microbial population of the soil. This aspect is often overlooked.

The ultimate objective of compost utilization in agriculture, horticulture, or land management is to increase plant growth or improve plant quality. However, another important function for compost is to improve soil physical conditions in order to enhance soil water conditions and reduce runoff and erosion, thereby reducing non-point-source pollution. Improving the soil-water relationship is extremely important in the revegetation of disturbed soil, such as previously mined areas or urban areas where the topsoil has been depleted. Compost is also used as an odor control product in biofiltration. In many cases compost is a substitute for other organic materials, such as peat. The use of compost is often more economical, since the cost of peat has increased as a result of transportation costs as well as mining costs.

The utilization and acceptance of compost depends on its quality. Quality is a function of the feedstocks and how the compost is prepared. The key quality criteria are the chemical, biological, and physical properties:

- Chemical
  - Plant nutrients
  - Trace elements
  - Salts
  - Organic compounds
- Biological
  - Stability
  - Maturity
  - Pathogens
  - Vector attraction
- Physical
  - Inerts
- Particle size

Knowledge and understanding of your product, and how best to utilize it for different applications, are essential to good marketing. One must know the product produced and how best to utilize it. The following are two examples. One compost marketer sold the product for use in avocados not knowing that the product had a salt content (high electrical conductivity) unsuitable for that plant. The result was that the grower incurred significant losses, which had to be compensated by the compost producer. Another instance where the use of unsuitable compost was used for a crop that could not tolerate the salt content resulted in total crop failure and a major lawsuit. It is often useful to provide the user with an analysis sheet and comments on the use and limitations of the compost.

The predominate uses of compost are:

- Horticulture
  - Ornamental and nursery crop production
  - Sod and turf production
  - Landscaping
- Agriculture
  - Vegetable production
  - Fruit trees
- Silviculture
- Public works
  - Disturbed land
  - Parks and cemeteries
  - Revegetation of public work activities
  - Highway—median strips and slopes
- Mined areas
- Miscellaneous areas for erosion and runoff control
- Plant disease control
- Odor control—biofiltration

In the sections below, some of the early discussion and examples are from my book *The Science of Composting* (Epstein, 1997). This material is supplemented with some information that has been subsequently reported.

Some suggested references on utilization of compost are:

Stoffella, P. J., and Kahn, B. A., eds., *Compost Utilization in Horticultural Cropping Systems*, Lewis Publishers, Boca Raton, FL, 2001
Epstein, E., *The Science of Composting*, CRC Press LLC, Boca Raton, FL, 1997
Brown, S., Angle, J. S., and Jacobs, L., eds., *Beneficial Co-utilization of Agricultural, Municipal, and Industrial By-products*, Kluwer Academic Publishers, Dordrecht, Netherlands, 1998

PROPERTIES OF COMPOST

Knowledge of the chemical, biological, and physical properties is very important for the proper utilization of compost. The chemical properties can and will affect plant growth. Furthermore, the content of major and minor plant nutrients could reduce the need for supplements, and hence affect the economics of production. Table 16.1 provides general information on the major plant nutrients found in various composts. The numbers are for illustrative purposes only. A compost producer should know his specific numbers. There is considerable variability in the plant nutrient content of compost as a result of the concentration of the elements in the feedstock and bulking agent used. Nitrogen is lost to the atmosphere as ammonia during composting, which reduces the final concentration in the compost. Nitrogen (N), phosphorus (P), and potassium (K) are plant nutrients listed as ingredients in fertilizer. In fertilizers, these are concentrated, and therefore need to be applied in relatively low rates. Compost contains these plant nutrients in considerably lower concentrations. What is often overlooked is that compost is applied at much higher rates in order to get the benefit of the organic matter. In doing so, we often provide a major portion or the entire N, P, and K needed for adequate plant growth. During composting, primarily some nitrogen is lost. Phosphorus does not change except for dilution with bulking agents, and potassium, although soluble, does not change except for dilution or leaching.

TABLE 16.1
Concentration of Total Nitrogen, Total Phosphorus, and Total Potassium in Several Composts

Compost	Nitrogen	Phosphorus	Potassium
Manure	1.5		
Biosolids	2.0–2.3	0.87–2.12	0.46–0.63
MSW	1.1–2.9	0.2–0.3	1.0–1.2
Yard waste	0.11	0.015	0.06
Poultry litter	1.01–1.26	1.05	0.68–0.83
Dairy manure	1.9	0.57	2.90

Chemical Properties

The three major chemical components of the feedstocks that could affect compost utilization and marketing are plant nutrients, heavy metals, and salts. The major plant nutrients, N, P, and K, vary with the feedstock as shown in Table 16.2.

Nitrogen in compost is predominantly in the organic form. Organic nitrogen is not soluble and will not leach through the soil or be available to plants. For nitrogen to be available to plants, it needs to be mineralized, i.e., converted to the inorganic soluble form. The inorganic forms of nitrogen are ammonium (NH_4^+) and nitrate (NO^3). Knowledge of the availability of nitrogen from compost is important in order to estimate the amount of N for crop growth. If the amount of available N from the compost is insufficient, then supplemental N is needed. If too much N is available from the compost, excess N beyond the crop's requirement can leach to groundwater.

Toxic organics, except for a specific compound such as chloropyralid, are generally not a problem. Many are decomposed during composting. These aspects were discussed in detail in the book *The Science of Composting* (Epstein, 1997) and will only be reviewed in this chapter. Most feedstocks are low in heavy metals. Plant materials and food waste are very low. Today, as a result of significant advances in wastewater technology, regulation, and reduction in the use of metals in construction, specifically lead and copper pipes, sewage sludge has much lower levels of heavy metals and trace elements (Epstein, 2002). Some trace elements can be toxic to plants, whereas others are essential (Epstein, 1997). Table 16.3 shows some of the trace elements in several composts. Several minor elements are often added as supplements in crop production, especially in the horticultural field. Again, depending on the feedstock, several minor elements may be available.

TABLE 16.2
Information on Some Plant Nutrients in Feedstocks Being Composted

Feedstock	Nitrogen %	Phosphorus % P	Potassium % K	Calcium %	Magnesium %	Reference
Dairy cattle	0.15–10.10	0.3–2.5	0.10– 6.50	0.25–2.80	0.11–0.71	Naylor et al., 1999
Beef cattle	1.90–7.80	0.41–2.60	0.44–4.20	0.53–5.00	0.29–0.56	Naylor et al., 1999
Poultry	1.3–14.5	0.15–4.00	0.55–5.40	0.71–14.90	0.30–1.30	Naylor et al., 1999
Swine	0.60–10.00	0.45–6.50	0.45–6.30	0.40–6.40	0.09–1.34	Naylor et al., 1999
Fish	2.04–3.94	0.56–4.67	0.06–0.23	3.00–11.20	0.04–1.93	Naylor et al., 1999
Paper mill sludge	1.75 (TKN)	0.021	0.025	0.414	0.045	Evanylo and Daniels, 1999
Sugarcane	2.42	1.81	0.10	7.62	0.43	Stoffella and Graetz, 2000

TABLE 16.3
Trace Elements in MSW, Yard Waste, Biosolids, and Food Waste Composts

Trace Element	MSW Compost mg/kg dry wt.	Yard Waste Compost mg/kg dry wt.	Biosolids Compost mg/kg dry wt.	Food Waste mg/kg dry wt.
As	1–4.8	1–12.8		3–4
B	34.5–113	0.2–76		21–109
Cd	1–13.2	0.04–0.81	3.6–16	<1
Cr	8.2–130	3.7–236	39–111	5–8
Cu	31–623	8–327	180–890	23–33
Hg	0.46–3.7	0.04–0.5	<0.01–3.5	<1
Ni	7–101	3.27–152	18–42	4–6
Pb	22–913	11.4–235	14–340	12–6
Zn	125–1,570	41.6–295	744–490	67–90

Note: NA = not available.
Major resources: Epstein (1997), Logan et al. (1999), Epstein et al. (2005).

Salt levels can be a significant problem in utilization of compost. Highly soluble salts in compost can result in detrimental plant growth or phytotoxicity. Food waste, biosolids, or sewage sludge could be two feedstocks that may contain high levels of soluble salts. One of the problems encountered in the foodwaste composting plant in South Korea is the high salt content of the incoming waste. Certain plants are more sensitive to salts than others (Bernstein, 1975). Soluble salt content, as measured by electrical conductivity (EC), can greatly affect germination and plant growth. Epstein (1997) indicated that electrical conductivity in excess of 5 m Scm^{-1} can result in phytotoxicity. Table 16.4 provides information on the sensitivity of crops to EC. In marketing compost it is essential for the marketer to know the potential use of the product. Table 16.5 shows the electrical conductivity of various compost materials determined by several researchers. What can a composter do if the product has a high electrical conductivity? If one knows that the feedstock has a high initial electrical conductivity, e.g., MSW or biosolids, it is best to incorporate a low-salt bulking agent or another feedstock having a low salt content. If, however, after composting, it has been determined that the product's intended use is for low-salt-tolerant crops, the salt content of the compost must be reduced. This can be done by leaching the salts out of the compost or diluting the compost with a low-salt compost product. Alternatively, market the compost to salt-tolerant plants.

Biological Properties

Compost from plants, animals, and humans, when incorporated into the soil, becomes part of the soil organic matter pool. The organic nitrogen undergoes mineralization by microorganisms. Ammonification converts the organic nitrogen to

TABLE 16.4
Crop Response to Salinity

Electrical Conductivity dS m^{-1} at 25°C	Crop Response
0–2	Negligible effects
2–4	Yields of sensitive crops affected
4–8	Yields of most crops affected
8–16	Salt-tolerant crops sustain yields
>16	Only very tolerant crops maintain yields

Source: Adapted from Bernstein, N. *Ann. Rev. Phytopathol.*, 13, 295–312, 1975.

TABLE 16.5
Electrical Conductivities of Various Composts

Composted Material	Conductivity dS/m	Reference
Yard waste compost	1.1–1.9	Campbell and Tripepi, 1991
Composted paper mill sludge	1.49–1.67	Evanylo and Daniels, 1999
Manure compost	3.62–3.65	Gonzalez et al., 1989
Urban refuse compost	7–12	Villar et al., 1993
MSW and biosolids compost	6.7	Ozores-Hampton et al., 2002
Pine bark	1.6	McLachlan et al., 2004
Paper mill sludge	2.2	McLachlan et al., 2004
Leaf and yard waste	3.5	McLachlan et al., 2004
MSW	12	McLachlan et al., 2004
Spent mushroom	12	McLachlan et al., 2004
Turkey litter	23.1	McLachlan et al., 2004
MSW	3.88	Hanay et al., 2004
Biosolids and yard trimmings	0.35–0.52	Klock-Moore, 1999

Source: Partially adapted from Agnew, J. M. and Leonard, J. J. *Compost Sci. Util.*, 11(3), 238–64, 2003.

ammonia (NH_3). Some of the NH_3 is converted to ammonium ions (NH_4^+) and is adsorbed on clay surfaces, where it is exchangeable with other cations or fixed in the clay structure. A major portion of NH_3 is nitrified by bacteria to nitrite (NO_2^-), which is very rapidly converted to nitrate (NO_3^+). Nitrate N is soluble and is available to plants, or if in excess of plant needs, it can leach with soil water into groundwaters. Denitrification under anaerobic soil conditions can convert NO_2^- and NO_3^+ to nitrogen (N_2) gas and be lost to the atmosphere.

It is important to know the mineralization rate of nitrogen from compost. This could vary with different composts since both the feedstock and bulking agent used could affect the mineralization rate. Epstein et al. (1978) determined the mineralization rates of compost from digested and raw biosolids compost. The highest amount of nitrogen mineralized occurred with the high application rate of nitrogen treatment of the digested biosolids compost. Less N was mineralized with raw biosolids compost than with digested biosolids compost. The percentage mineralized over a fifteen-week period was in the range of 7 to 10% for the digested biosolids compost and 4 to 5% for the raw biosolids compost. Based on all the studies reviewed, the rate of nitrogen mineralization from cured compost usually ranges between 7 and 10% over a twenty- to thirty-five-week period. Nitrogen mineralization is rapid initially and then decreases and remains fairly constant. There are laboratories that can provide more exact mineralization rates for specific composts.

The important biological properties to the user are stability, maturity, and pathogens. Stability is achieved through proper composting, especially during curing. The destruction of pathogens during composting is discussed in Chapter 13. Pathogens, stability, and maturity are discussed in detail in Chapter 5 in the book *The Science of Composting* (Epstein, 1997). Stability is a stage in the decomposition of organic matter and is a function of biological activity. Stability of compost is achieved through proper curing. A stable product, when incorporated into the soil, will not decompose rapidly and utilize nitrogen required for plant growth. The most accepted method of measuring stability is respiration. Table 16.6 illustrates values indicating the degree of stabilization based on respiration.

If compost is not stable, it will continue to decompose rapidly in the soil. During that period, nitrogen in the soil, which can be available for plants, is utilized by microorganisms. As a result, nitrogen deficiency to plants can occur. When a compost that has a high C:N ratio is added to soil, the microbial population competes with plants for soil nitrogen. In this case, plants typically exhibit chlorosis, yellowing

TABLE 16.6
Compost Stability Index Based on Carbon Dioxide Evolution

Respiration Rate mg CO_2-C/g compost C-day	Rating	Characteristics
<2	Very stable	Well cured; no malodors; earthy odor
2–5	Stable	Cured compost; minimal impact on soil dynamics
5–10	Moderately stable	Uncured compost; some malodor potential; addition to soil may immobilize N; high phytotoxicity potential; not recommended for growing compost from seed
10–20	Unstable compost	Very immature compost; high malodor and phytotoxicity potential; not recommended for growing plants from seed
>20	Very unstable compost	Extremely unstable material; very high malodors and phytotoxicity potential; not recommended for use

of the leaves, indicating a nitrogen deficiency. If organic materials having low C:N ratios are added to soil, ammonia can be released, which can cause phytotoxicity (Dowdy et al., 1976). The decrease in oxygen in soils, as a result of microbial respiration, can result in reduced oxygen levels or anaerobic conditions, which will not only influence plant growth, but also affect soil chemical species, heavy metal solubility, and uptake of nutrients.

Maturity is an organochemical condition of the compost that indicates the presence or lack of phytotoxic organic acids. This distinguishes between the organic and phytotoxic effects due to inorganic chemicals and salinity. The organochemical phytotoxic effects are primarily attributed to fatty acid formation. However, there are other organic compounds present in immature compost that could result in phytotoxicity. Maturity of compost is achieved through the destruction of fatty acids, which primarily occur during proper curing. Maturity is measured by seed germination.

Another important aspect of compost is its potential for plant disease suppression (Epstein, 1997). This value needs to be stressed in certain uses for horticultural and turf grass.

Physical Properties

The most important physical properties affecting the utilization and marketing of compost are physical contaminants and particle size. Many European countries and Canadian provinces regulate the level of contaminants. Several states have incorporated regulations of physical contaminants. There are no U.S. federal regulations. Generally MSW is the most common feedstock that results in the presence of inerts, such as glass, plastics, and stones. These are generally removed by screening and air separation. Particle size is important, depending on the use of the compost. Compost used primarily as a mulch can have large particles, whereas a compost to be used for grass, sod, or turf usually demands a very small particle size.

A clean compost devoid of stones, glass, paper, metal pieces, and other contaminants is essential for utilization and marketing. Generally, composting of homogeneous feedstocks such as biosolids, food waste, and clean yard waste does not present a problem. Other than particle size, which is achieved through screening, no further refining is needed. MSW is very heterogeneous, and the compost, prior to refining, can be very contaminated. Manure is generally physically clean. However, stainless needles used for injection of antibiotics and other items have been found in the compost and can be a problem for removal.

Compost primarily is a soil conditioner. It improves several of the soil physical properties essential for plant growth, water, and air movement through the soil. What is meant by stating "it enhances the soil physical properties?" The important soil physical properties are soil structure, soil water, and soil aeration. Soil structure refers to soil aggregation. Organic matter is an important component of soil aggregates. Proper soil structure is important for water movement and soil aeration, as well as enhanced plant root development and growth. In sandy soils, the organic matter from compost provides greater soil water retention. The increased soil water provides for greater water availability for plants and reduces the need for irrigation.

In heavy soils containing large amounts of clay, the organic matter "breaks up" the clay particles. This enhances water movement, and therefore water infiltrates the soil and reduces runoff and erosion. Compost can significantly reduce erosion from highway slopes, home construction, and public works.

Figure 16.1 shows the effect of compost on soil aggregation. Tilling compost into a heavy soil, i.e., a soil containing considerable clay and tending to be compacted, loosens the soil, provides aggregation, and allows water infiltration.

In Figure 16.2 the use of compost improved the structure of a heavy soil, and thus provided for better root development, as shown by the corn plant on the left and the field on the right. Better water retention and availability also occurred. Notice the darker soil color because of the compost application. The darker soil also increases soil temperature in the spring and enhances seed germination.

Compost is also used as mulch, in contrast to incorporating the compost into the soil. Mulch serves two purposes. One purpose is to reduce the impact of raindrops

FIGURE 16.1 Effect of compost on soil physical properties.

FIGURE 16.2 The effect of compost on plant root development and resulting increase in corn growth.

on the soil. This prevents the soil surface from sealing and increased runoff. The second purpose is to reduce evaporation, and thus conserve moisture, making it more available for plants.

UTILIZATION OF COMPOST

An excellent monograph on this subject is by Stoffella and Kahn (2001). Another good document is by CIWMB (2007).

A brief review of the uses of compost is presented in the following sections, primarily to indicate to the reader the variability and extensiveness of the use of compost.

General Uses

The primary use of compost is as a soil amendment to improve the soil physical properties. These involve:

- Improved soil structure and porosity
  - Improved soil aeration
  - Improved soil aggregation
  - Improved water relations of plants
- Improved water-holding capacity of sandy soils
- Improved water infiltration and drainage in heavy soils
- Increased organic matter content
  - Increased cation exchange capacity
- Increased activity and diversity of soil microorganisms

The improvement in soil structure and porosity results from the organic matter effect on soil particles or for aggregates. This improvement in soil structure provides for better soil aeration. Thus, plant roots can proliferate and develop. The results are better utilization of plant nutrients and water. Sandy soils have little or no structure. The added organic matter, through the use of compost, binds the sand particles to provide better structure to the sandy soil. The soil is then better able to retain plant nutrients and water. The addition of compost to heavy, clay soils improves the soil structure for better water infiltration, permeability, and drainage. These enhancements can reduce soil erosion and runoff.

Horticulture

The greatest use of compost is probably in horticulture, although agriculture, particularly organic farming, could be the greatest user in the future. The horticultural industry utilizes vast quantities of organic materials in the production of marketable plants. The high value of the crops grown enables the user to invest in high-quality products. In recent years the cost of natural organic materials such as peat has increased. The container-grown plant industry uses a soilless potting media consisting of 75% or more of peat and pine bark (Shiralipour et al., 1992). Many different media, including rice hulls, pine bark, hardwood bark, and other material, have been utilized successfully as a replacement for peat moss. Compost can be an excellent substitute for peat or other media being used. The two most important requirements by the industry are quality and consistency.

Producers of greenhouse and nursery crops are ideal users of waste composts. Under these conditions there is less potential health risk from pathogens or heavy metals. However, persons working in the container industry have expressed concern about health when using compost prepared with biosolids and other feedstocks. Meeting the federal 503 regulations assures workers of the safety of the compost from pathogens. The presence of glass shards or other inerts in MSW compost can pose a concern to the user's safety. There is equipment to remove inerts (glass, plastics, sand, etc.). It is important for the compost industry to maintain high standards for compost quality.

Ornamentals

In the book *The Science of Composting* (Epstein, 1997), I referenced many early studies. During the 1970s considerable research was conducted on the utilization of compost. Composting on a large scale was relatively new, especially biosolids composting. Consequently, there was a rush to evaluate the benefits of compost. Principally, the Universities of Maryland, Florida (Gainesville), Ohio, and Cornell conducted studies in horticulture. Table 16.7 lists some studies after 1977. Data on flowering plants are shown in Table 16.8.

It is clear from these and previous studies that individual plant species may respond very differently to different types of compost. This response appears to be due to sensitivity to soluble salts, moisture conditions, bulk density, and specific elements such as boron (B). Boron was at times high in MSW compost. Many of these conditions can be easily modified prior to using the compost. Soluble salt levels, as well as boron and other soluble elements, can be leached prior to planting or diluted with compost, sawdust, and other material having very low salt contents. Moisture, bulk density, and other physical parameters can be modified during preparation of the planting media.

Sod Production, Turf Grass, and Lawn Establishment

The production of sod represents one of the best potential uses for compost. Harvested sod removes a layer of topsoil and eventually depletes the entire topsoil. The use of compost for turf grass production for home lawns, parks, athletic fields, cemeteries, or institutional grounds can result in improvement of the soil physical properties, while adding nutrients and organic matter. Compost can be used in turf grass production as (1) a soil amendment for the establishment of turf grass, (2) a fertilizer source for maintenance of established turf grass, and (3) a soil amendment or growth medium for commercial sod production (Hornick et al., 1979). Figure 16.3 shows studies on the use of compost for turf production, and Figures 16.4 and 16.5 show the use of compost in developing the White House lawn.

TABLE 16.7
Utilization of Compost on Ornamentals during 1977 to 2008

Ornamentals	Compost Media	Details	References
Tulip poplar	Bio/WC	2.5, 5, and 10 cm thickness	Gouin and Walker, 1977
Dogwood	Bio/WC	2.5, 5, and 10 cm thickness	Gouin and Walker, 1977
Norway spruce	Bio/WC	2.5, 5, and 10 cm thickness	Gouin, 1977
White pine	Bio/WC	2.5, 5, and 10 cm thickness	Gouin, 1977
Dogwood, forsythia, ninebark, weigela	Paper mill sludge	0, 20, 40, 60% by vol. with bark, topsoil, and sand	Chong, 2003
Coralberry, snowberry, viburnum, lilac, osier, sandcherry, euonymous	MSW	0, 15, 30, 45, 60, 70% amended with perlite or sphagnum peat	Chong, 1999
Baby doll, dieffenbachia, dracaena, cane	MSW/bio., YT	Various percentages with sphagnum peat or pine bark	Chen et al., 2002
Silverleaf dogwood, red-osier dogwood, deutziaa, ninebark	Wax corrugated cardboard	25 to 50% compost in rooting medium	Raymond et al., 1998
Dwarf oleander, orange-jessamine	Paper mill, MSW	100%	Fitzpatrick, 1989
West Indian mahogany, pink tabebula, pigeon-plum, key lime, schiflera	Bio	80%	Fitzpatrick, 1985

Note: Bio = biosolids, MSW = municipal solid waste.

AGRICULTURE

Although agriculture could represent the greatest potential for the utilization of compost, it currently does not use much compost. The exception is organic farming, which is growing. There have been numerous articles on the use of compost for agronomic crops.

Table 16.9 provides a summary of the literature from 1998 to 2008. Maynard and Hill (2000) reported that after three years of compost application, yields of three Spanish onion cultivars were significantly greater than yields in plots not receiving compost.

TABLE 16.8
Several 1998–2008 Citations Regarding Use of Compost for Flowering Plants

Flowering Plant	Composting Media	Details	Reference
Marigolds	Paper mill sludge	Supplements	Evanylo and Daniels, 1999
Hedera helix L.	Straw	Various N sources	Kresten Jensen et al., 2001
Cordyline terminalis 'Baby doll'	MSW, bio yard trimmings	Comparison with other media	Chen et al., 2002
Dieffenbachia maculate 'Camille'	MSW, bio yard trimmings	Comparison with other media	Chen et al., 2002
Dracaena fragans 'Massangeana'	MSW, bio yard trimmings	Comparison with other media	Chen et al., 2002
Various ornamental plants	Bio	24 plants	Bugbee, 2002
Impatiens, salvia	Bio yard trimmings	Five substrates	Klock-Moore, 1999

Note: Bio = biosolids, MSW = municipal solid waste.

FIGURE 16.3 Compost use in turf production. The light grass did not get any compost but normal fertilizer. Different shades of green indicate different application levels of compost.

FIGURE 16.4 Applying compost to the White House lawn.

FIGURE 16.5 An established lawn at the White House following compost application.

TABLE 16.9
Some Studies from 1998 to 2008 as Related to Agricultural Use of Compost

Crop	Compost Media	Comments	References
Spanish onions	Leaf compost	Fertilizer addition	Maynard and Hill, 2000
Organic wheat	Dairy manure	Five rates	Stukenholtz et al., 2002
Tomatoes	Leaf compost	Fertilizer additions	Maynard, 2000
Wheat, barley, canola	MSW/biosolids	Fertilizer additions, metal uptake	Zhang et al., 2000
Forage maize	Greenwaste	Fertilizer additions, organic mater	Parkinson et al., 1999
Tomato transplants	Biosolids/yard trimmings	Combination with other amendments	Ozores-Hampton et al., 1999

Forestry and Land Reclamation

The use of compost in forest nurseries is an excellent market. The literature on forest application is very sparse. Early research was conducted by Frank Gouin, PhD, at the University of Maryland. Subsequent research was conducted by the University of Florida.

Compost can be very effectively used in mine land reclamation, as well as other disturbed soils. These soils are low in organic matter and fertility. Compost applied to these soils adds organic matter, increases fertility, and improves soil physical properties, thereby assisting in revegetation of disturbed soils. This is shown in Figure 16.6. Most of the published research on utilization of compost for mine and disturbed soil reclamation was produced in the 1970s.

FIGURE 16.6 The use of compost for coal mine spoil reclamation.

In an attempt to reclaim desert land and make it productive, MSW compost was applied, along with irrigation, in a study in Yuma, Arizona. This effort is shown in Figure 16.7: a one-time application of compost-established crops. This clearly shows that poor soils in many parts of the world can be made productive. Human and animal waste, when properly composted to destroy pathogens, can be an excellent material for soil amelioration.

One of the most elaborate reclamation sites using compost was Constitution Gardens in Washington, DC (Figures 16.8 and 16.9). The project was initiated

FIGURE 16.7 Studies in Yuma, Arizona, use compost to establish crops on desert soil.

FIGURE 16.8 Use of compost in Constitution Garden in Washington, DC.

FIGURE 16.9 Compost applied to Constitution Garden in Washington, DC.

prior to the two hundredth birthday of the United States (Patterson, 1975; Cook et al., 1979). Considerable topsoil would have been required for the forty-two-acre project. The Ecological Services Laboratory of the National Capital Parks, U.S. Department of Interior, felt that biosolids compost produced at the U.S. Department of Agriculture's Beltsville project mixed with leaf mold could be used for soil modification of the existing soil. Today, Constitution Garden is thriving and is an excellent example of the use of urban wastes. Compost was also used at the Statue of Liberty State Park in New Jersey, as shown in Figures 16.10 and 16.11.

Figure 16.10 shows the poor soil at Statue of Liberty State Park in New Jersey, prior to applying compost. Compost improved the soil physical properties and a good lawn was established, as shown in Figure 16.11.

Public Works

The use of compost for public works is very effective. In Figures 16.12 and 16.13, the application of compost on a road slope established grass and prevented run-off and erosion. In some states compost is very effective in enhancing highway median strips. State highway departments need to specify compost in the construction and maintenance of their highway system. This, in turn, would encourage communities to establish composting facilities, as they will have a market for their products.

Compost can be effectively used in flowerbeds and other public areas to enhance the beauty of the community. Durham, New Hampshire, used compost throughout the city and placed small signs near the flowerbeds stating that this was one use for the tax dollars.

FIGURE 16.10 Poor soil at Liberty State Park, New Jersey. Prior to applying compost.

FIGURE 16.11 An established lawn over soil enriched with compost at Liberty State Park, New Jersey.

COMPOST MARKETING

One of the most important aspects of product marketing is knowledge of the product's assets and limitations. This requires a laboratory analysis.

As was pointed out earlier, compost is organic matter that enhances the soil physical properties. Is it not a fertilizer? Although it contains low levels of plant nutrients, it is not classified or regarded as a fertilizer, unless additional macronutrients are

FIGURE 16.12 Bare soil causing erosion and runoff near a high school in Maryland.

FIGURE 16.13 Use of compost for highway erosion and runoff control at a high school in Maryland.

supplied. However, as indicated, it can contain low levels of major nutrients (nitrogen, phosphorus, and potassium) and minor elements, such as iron, manganese, boron, copper, and others (Epstein, 1997). However, since the soil benefits by the addition of large amounts of organic matter, significant amounts of major and minor elements are provided. When a compost product contains 1.5% nitrogen, and then if 0.9078 tonne (1 ton, 907 kg, 2,000 lb) is applied to the soil (which is a very low a amount), 18.1 kg (20 lb) of nitrogen is provided for plant growth. Furthermore, there is a significant value and benefit to the minor elements in the compost. Therefore, it is important to analyze the compost and know its benefits. Compost can contain salts. Depending on the salinity level, it can be detrimental to some plants. This is indicated by the electrical conductivity (EC) value. MSW and biosolids compost may have high EC values, which can restrict their use (Epstein, 1997; Stoffella and Kahn, 2001).

The marketer must be aware of these properties of compost. The key to successful marketing is:

- Know your product.
  - How did you make it?
  - What is it made of? What are its properties?
  - Where did you make it? Know your company and process.
  - Why did you make it? It reduces organic material to a landfill. It is the ecological and "green" thing to do. The compost reduces greenhouse gases by adding carbon to the soil.
- Know your competition. What products do they produce? What is the quality of the products? Which ones have a high demand? What are the products selling for?
- Develop a marketing plan and be prepared to make modifications as the market changes.

CONCLUSION

The potential utilization and marketing are important aspects in the planning, design, and operations of a composting facility. Consideration must be given to transportation, distance to markets, type of markets, demand for products, and greatest potential remuneration.

Currently, the horticultural market is the best market for compost use. Near urban areas, the use of compost in turf production can be excellent.

The use of compost on disturbed lands such as parks, highways, gravel pits, and mine reclamation has been very effective. It can be used for highway reclamation projects. This can result in savings to municipalities and other agencies.

The use and sale of compost will lower operational costs for facilities. If tipping fees or avoided costs can pay for a facility's operation, then the sale of compost can result in profit.

REFERENCES

Agnew, J. M., and Leonard, J. J. 2003. The physical properties of compost. *Compost Sci. Util.* 11(3):238–64.

Bernstein, N. 1975. Effects of salinity and sodicity on plant growth. *Ann. Rev. Phytopathol.* 13:295–312.

Bugbee, G. J. 2002. Growth of ornamental plants in container media amended with biosolids compost. *Compost Sci. Util.* 10(2):92–98.

Campbell, A. G., and Tripepi, R. R. 1991. Composting of wood and yard wastes. *Forest Products J.* 41(3):55–57.
Chen, H., McConnell, D. B., Robinson, C. A., Caldwell, R. D., and Huang, Y. 2002. Production and interior performances of tropical ornamental foliage plants grown in container substrates amended with compost. *Compost Sci. Util.* 10(3):217–23.
Chong, C. 1999. Rooting of deciduous wood stem cuttings in peat- and perlite-amended MSW compost media. *Compost Sci. Util.* 7(4):6–14.
Chong, C. 2003. Paper mill waste mixed with compost and other ingredients as container nursery substrates. *Compost Sci. Util.* 11(1):16–26.
Cook, R. N., Patterson, J. C., and Short, J. R. 1979. Compost saves money in parkland restoration. *Compost Sci. Land Util.* 20(2):43–44.
Dowdy, R. H., Larson, R. E., and Epstein, E. 1976. *Sewage sludge and effluent use in agriculture.* Ankeny, IA: Land Application of Waste Material, Soil Conservation Society of America.
Epstein, E. 1997. *The science of composting.* Boca Raton, FL: CRC Press LLC.
Epstein, E. 2002. *Land application of sewage sludge and biosolids.* Boca Raton, FL: Lewis Publishers.
Epstein, E., Keane, D. B., Mesinger, J. J., and Legg, J. O. 1978. Mineralization of nitrogen from sewage sludge and sludge compost. *J. Environ. Qual.* 7, 217–221.
Evanylo, G. K., and Daniels, W. L. 1999. Paper mill sludge composting and compost utilization. *Compost Sci. Util.* 7(2):30–39.
Fitzpatrick, G. 1985. Container production of tropical trees using sewage effluent, incinerator ash, and sludge compost. *J. Environ. Hort.* 3(3):123–25.
Fitzpatrick, G. 1989. Solid waste compost as growing media. *BioCycle* 30(9):62–64.
Gonzalez, J. L., Medina, M., and Benitez, I. C. 1989. Slurry composting options. *BioCycle* 30(7):53–55.
Gouin, F. R. 1977. Confer tree seedling response to nursery soil amended with composted sewage sludge. *Hortscience* 12:341–42.
Gouin, F. R., and Walker, J. M. 1977. Deciduous tree seedling response to nursery soil amended with composted sewage sludge. *Hortscience* 12:45–47.
Hanay, A., Buyuksonmez, F., Kiziloglu, F., and Canbolat, M. Y. 2004. Reclamation of saline-sodic soils with gypsum and MSW compost. *Compost Sci. Util.* 12(2):175–79.
Hornick, S. B., Murray, J. J., Chaney, R. L., Sikora, L. J., Parr, J. F., Burge, W. D., Willson, G. B., and Tester, C. F. 1979. *Use of sewage sludge compost for soil improvement and plant growth.* Beltsville, MD: U.S. Department of Agriculture, Science, and Education Administration.
Klock-Moore, K. 1999. Growth of impatiens ‘Accent Orange’ in two compost products. *Compost Sci. Util.* 7(1):58–62.
Kresten Jensen, H. E., Leth, M., and Lonsmann, J. J. 2001. Growth of *Hedera helix* L. container plants in compost substrates made with *Miscanthus ogiformis* Honda straw and various N-sources. *Compost Sci. Util.* 9(3):206–14.
Logan, T. J., Henry, C. L., Schnoor, J. L., Overcash, M., and McAvoy, D. C. 1999. An assessment of health and environmental risks of trace elements and toxic organics in land-applied municipal solid waste compost. *Compost Sci. Util.* 7(3):38–53.
Maynard, A. A. 2000. Applying leaf compost to reduce fertilizer use in tomato production. *Compost Sci. Util.* 8(3):203–9.
Maynard, A. A., and Hill, D. E. 2000. Cumulative effect of leaf compost on yield and size distribution in onions. *Compost Sci. Util.* 8(1):12–18.
McLachlan, K. L., Chong, C., Voroney, R. P., Liu, H., and Holbein, B. E. 2004. Variability of soluble salts using different extraction methods on compost and other substrates. *Compost Sci. Util.* 12(2):180–84.

Naylor, S. J., Moccia, R. D., and Durant, G. M. 1999. The chemical composition of settleable solid fish waste (manure) from commercial rainbow trout farms in Ontario, Canada. *North Am. J. Agric.* 61:21–26.

Ozores-Hampton, M., Obreza, T. A., Stoffella, P. J., and Fitzpatrick, G. 2002. Immature compost suppresses weed growth under greenhouse conditions. *Compost Sci. Util.* 10(2):105–13.

Ozores-Hampton, M., Vavrina, C. S., and Obreza, T. A. 1999. Yard trimmings-biosolids compost: Possible alternative to sphagnum peat moss in tomato transplant production. *Compost Sci. Util.* 7(4):42–49.

Parkinson, R. J., Fuller, M. P., and Groenhof, A. C. 1999. An evaluation of greenwaste compost for the production of forage maize (*Zea mays* L.). *Compost Sci. Util.* 7(1):72–80.

Patterson, J. C. 1975. Enrichment of urban soil with composted sludge and leaf mold—Constitution Gardens. *Compost Sci.* 16(30):18–21.

Raymond, D. A., Chong, C., and Voroney, R. P. 1998. Response of four container grown woody ornamentals to immature composted media derived from waxed corrugated cardboard. *Compost Sci. Util.* 6(2):67–74.

Shiralipour, A., McConnell, D. B., and Smith, W. H. 1992. Uses and benefits of MSW compost: A review and an assessment. *Biomass Bioenergy* 3(3–4):267–79.

Stoffella, P. J., and Graetz, D. A. 2000. Utilization of sugarcane compost as a soil amendment in a tomato production system. *Compost Sci. Util.* 8(3):210–14.

Stoffella, P. J., and Kahn, B. A. 2001. *Compost utilization in horticultural cropping systems.* Boca Raton, FL: Lewis Publishers.

Stukenholtz, P. D., Koenig, R. T., Hole, D. J., and Miller, B. E. 2002. Partitioning the nutrient and nonnutrient contributions of compost to dryland-organic wheat. *Compost Sci. Util.* 10(3):238–43.

Villar, M. C., Beloso, M. C., Acea, M. J., Cabaneiro, A., Gonzalez-Prieto, S. J., Diaz-Ravina, M., and Carballas, T. 1993. Physical and chemical characterization of four composted urban refuses. *Bioresource Technol.* 45:105–13.

Zhang, M., Heaney, D., Solberg, E., and Heriquez, B. 2000. The effect of MSW compost on metal uptake and yield of wheat, barley, and canola in less productive farming soils of Alberta. *Compost Sci. Util.* 8(3):224–35.

Index

www.ingramcontent.com/pod-product-compliance
Ingram Content Group UK Ltd.
Pitfield, Milton Keynes, MK11 3LW, UK
UKHW022216190926
475702UK00003B/380
* 9 7 8 1 4 3 9 8 4 5 3 1 8 *